Wissenschaftliche Reihe Fahrzeugtechnik Universität Stuttgart

Reihe herausgegeben von

André Casal Kulzer, Stuttgart, Deutschland

Hans-Christian Reuss, Stuttgart, Deutschland

Andreas Wagner, Stuttgart, Deutschland

Das Institut für Fahrzeugtechnik Stuttgart (IFS) an der Universität Stuttgart forscht interdisziplinär sowie technologieoffen an modernen und zukunftsorientierten Fahrzeugkonzepten. In enger Zusammenarbeit mit Partnern aus Industrie und Wissenschaft entstehen neue Lösungen für die Mobilität der Zukunft. Das Institut gliedert sich in drei spezialisierte Lehrstühle, die gemeinsam das gesamte Spektrum der Fahrzeugtechnik abdecken: Der **Lehrstuhl für Fahrzeugantriebssysteme** widmet sich der Forschung nachhaltiger Antriebslösungen für künftige Mobilitätskonzepte. Im Fokus stehen alternative, elektrische sowie hybride Antriebssysteme und deren Komponenten – einschließlich der Nutzung nachhaltiger Energieträger wie Wasserstoff, synthetischer Kraftstoffe und Batterien. Der **Lehrstuhl für Kraftfahrzeugmechatronik** beschäftigt sich mit vernetzten, intelligenten und adaptiven Fahrzeugen. Im Zentrum stehen Fragestellungen zum Automatisierten und Vernetzten Fahren, Diagnose, Ladetechnologien, verteilte Systeme sowie softwarebasierte Fahrzeugfunktionen. Der **Lehrstuhl für Kraftfahrwesen** erforscht die physikalischen Grundlagen der Auslegung zukünftiger Fahrzeugkonzepte. Im Mittelpunkt stehen die Bereiche Aerodynamik, Windkanaltechnik, Akustik/NVH, Fahrzeugdynamik, Reifenmanagement und Thermomanagement Gesamtfahrzeug. Das IFS verfügt über eine vielfältige und hochmoderne Forschungsinfrastruktur, die realitätsnahe Untersuchungen vom Einzelbauteil bis zum Gesamtfahrzeug ermöglicht. Besonders hervorzuheben sind der Multikonfigurations- und Antriebsprüfstand für komplexe Antriebskonzepte, der Stuttgarter Fahrsimulator zur Untersuchung menschlichen Fahrverhaltens, der Aeroakustik-Fahrzeugwindkanal für akustische und strömungstechnische Fragestellungen sowie der Thermowindkanal zur Analyse thermischer Prozesse im Gesamtfahrzeug. Die wissenschaftliche Reihe „Fahrzeugtechnik Universität Stuttgart" dokumentiert die im Rahmen von Promotionen am IFS entstandene Beiträge zur Mobilität der Zukunft und zeigt deren thematische Vielfalt, methodische Tiefe und Praxisrelevanz.

Reihe herausgegeben von

Prof. Dr.-Ing. André Casal Kulzer
Lehrstuhl Fahrzeugantriebssysteme
Institut für Fahrzeugtechnik Stuttgart
Universität Stuttgart
Stuttgart, Deutschland

Prof. Dr.-Ing. Hans-Christian Reuss
Lehrstuhl Kraftfahrzeugmechatronik
Institut für Fahrzeugtechnik Stuttgart
Universität Stuttgart
Stuttgart, Deutschland

Prof Dr.-Ing. Andreas Wagner
Lehrstuhl Kraftfahrwesen
Institut für Fahrzeugtechnik Stuttgart
Universität Stuttgart
Stuttgart, Deutschland

Alia Salah

Model-Based Fault Diagnosis and Fault-Tolerant Control

An Approach for Automotive Electric Machines

Springer Vieweg

Alia Salah
Institute of Automotive Engineering
(IFS), Chair in Automotive Mechatronics
University of Stuttgart
Stuttgart, Germany

Zugl.: Dissertation Universität Stuttgart, 2025
D93

ISSN 2567-0042 ISSN 2567-0352 (electronic)
Wissenschaftliche Reihe Fahrzeugtechnik Universität Stuttgart
ISBN 978-3-658-50107-5 ISBN 978-3-658-50108-2 (eBook)
https://doi.org/10.1007/978-3-658-50108-2

This Springer Vieweg imprint is published by the registered company Springer Fachmedien Wiesbaden GmbH, part of Springer Nature.
The registered company address is: Abraham-Lincoln-Str. 46, 65189 Wiesbaden, Germany

Acknowledgment and Dedication

This Dissertation is created during my work as a research scientist at the Research Institute of Automotive Engineering and Vehicle Engines (FKFS) .

My heartfull appreciation goes to Prof. Dr.-Ing. Hans-Christian Reuss for his guidance and unwavering support throughout the journey of this thesis.

I am grateful to the member of my thesis committee, Prof. Dr.-Ing. Nejila Parspour for her constructive feedback and essential suggestions that enhanced the quality of my work.

I would like to express my appreciation to Dr. Tansel Yucelen from the University of South Florida for his generosity and devotion in providing detailed lectures and materials on model reference adaptive control which played a pivotal role in shaping and refining the presented fault-tolerant control approach.

Gratitude fills my heart for the invaluable contributions and for the unwavering support and guidance of my former colleague and husband, Dr. Omar Abu Mohareb. For His expertise and patience, for the countless discussion hours, for his constant support and unlimited belief in me all of that and more, were the cornerstone of this academic endeavor.

Acknowledgment also goes to my colleagues for their support and collaboration.

To my beloved daughters Ghaida and Hanna who filled my journey with unlimited love and all the beautiful things in life. Their unconditional love motivated me to persist and keep advancing.

Finally, I dedicate this dissertation to my father, Jamal Salah, in honor of his memory and to realize his aspirations.

Stuttgart Alia Salah

Contents

List of Figures

List of Tables

Abbreviations

AFTC	Active Fault Tolerant Control
AI	Artificial Intelligence
BEV	Battery Electric Vehicle
CCA	Cold Cranking Amps
CO	Corrosion Fault
D	Diode Fault
DF	Diode Fault
DT	Duty Cycle
DTC	Diagnostic Trouble Codes
DTw	Digital Twin
DTwI	Digital-Twin Instance
DTwP	Digital-Twin Prototype
E-machine	Electric Machine
ECU	Electronics Control Unit
EESM	Externally Excited Synchronous Machine
EPGS	Electric Power Generation and Storage
EV/HEV	Electric Vehicles/ Hybrid Electric Vehicles
FA	Flaking Fault
FD	Fault Detection
FDD	Fault Detection and Diagnosis
FDI	Fault Detection and Identification
FI	Fault Identification
FKFS	Forschungsinstitut für Kraftfahrwesen und Fahrzeugmotoren Stuttgart
FNR	False Negative Rate
FOC	Field Oriented Control

FTC	Fault-Tolerant Control
GCD	Greatest Common Divisor
GS	Gain Scheduling
HE	Healthy
HIL	Hardware-in-the-Loop
ICE	Internal Combustion Engine
IFAC	International Federation of Automatic Control
IoT	Internet of Things
IPMSM	Internal Permanent Magnet Synchronous Machine
IVK	Institut für Verbrennungsmotoren und Kraftfahrwesen
KA	Corrosion Fault
LCM	Least Common Multiple
ML	Machine Learning
MPC	Model Predictive Control
MRAC	Model Reference Adaptive Control
MTPA	Maximum Torque per Ampere
NB	Naive Bayes
NN	Neural Networks
OBD	On-Board Diagnostic
OBDII	Off-Board Diagnostic
OEM	Original Equipment Manufacturer
OTA-Update	Over-the-Air-Update
PFTC	Passive Fault Tolerant Control
PI	Proportional Integral
PMSM	Permanent Magnet Synchronous Machine
PWM	Pulse Width Modulation

QD	Quadratic Discriminant
SDV	Software-Defined Vehicle
SMC	Sliding Mode Control
SVM	Support Vector Machine
TPR	True Positive Rate
UDS	Off-Board Diagnostic
WS	Wire String Fault

Nomenclature

	Greek Letters	
α	Alternator Model Parameter	-
β	Alternator Model Parameter	Ω^{-1}
η	Alternator Model Parameter	$\frac{V}{\Omega \cdot rad \cdot s^{-1}}$
η_{eff}	Efficiency of Electric Machine	%
γ	Alternator Model Parameter	-
γ_a	Adaptation Rate	-
κ	Alternator Model Parameter	-
Λ	Contro Ineffectiveness Matrix	-
λ	Alternator Model Parameter	Ω^{-1}
$\lambda_a, \lambda_b, \lambda_c$	Flux Linkage of the Phases	Wb
λ_d	Direct Axis Flux Linkage	Wb
λ_f	Flux Linkage of Field	Wb
λ_{pm}	Permanent Magnet Flux Linkage	Wb
λ_q	Quadrature Axis Flux Linkage	Wb
ω_e	Electrical Speed	rad/s
ω_a	Adaptive Rotational Speed Signal	rpm
σ	In-parameter State-dependent Matched Uncertainty	-
θ_e	Alternator Phase Angle	rad
θ_m	Angular Displacement of Alternator	rad
Θ_p	Estimated Parameters Set of the Alternator	
Θ_{pf}	Parameters Vector of the System with Fault	-
ξ	Aggregated State Vector	-

	Indices
F	System-with-Fault
H	Healthy System
i	Current
n	Nominal

ref	Reference	
	Latin Letters	
A	State Matrix	-
A_r	Aggregated State Matrix	-
B	Input Matrix	-
B_f	Inertia Coefficient	kgm^2/s
B_r	Aggregated Input Matrix	-
$C_{v_{ref}}$	On-Broad Reference Voltage	V
DT	Duty Cycle	%
DT_a	Adaptive Duty Cycle	%
E	Output Matrix	-
e	Tracking Error	-
E_a, E_b, E_c	Induced Electromotive Forces	V
E_f	Induced Electromotive Force of the Field	V
e_{I_dc}	Deviation Error in Alternator Current	A
e_{I_f}	Deviation Error in Field Current	A
Er	Deviation Error	-
f_a, f_b, f_c	Parameters of the 3-Phase Passive Rectifier	-
g_a, g_b, g_c	Conduction State of Diode	-
I_a, I_b, I_c	AC Phase Currents	A
I_{Bat}	Battery Current	A
I_d	Direct Axis Current	A
L_d	Direct Axis Inductance	H
I_{dc}	Alternator DC Current	A
I_f	Alternator Field Current	A
V_f	Alternator Field Voltage	V
I_q	Quadrature Axis Current	A
J	Inertia Coefficient	kgm^2
K_{i_1}, K_{i_2}	Parameters of the Motor FOC Controller	-
L_{ab}, L_{ba}	Stator-Stator Mutual Inductance	H
L_a, L_b, L_c	Self Inductance of the Stator Windings	H
L_{ac}, L_{ca}		H
L_{bc}, L_{cb}		H
L_{af}, L_{fa}	Stator-Rotor Mutual Inductance	H

L_{bf}, L_{fb}	Stator-Rotor Mutual Inductance	H
L_{cf}, L_{fc}	Stator-Rotor Mutual Inductance	H
L_f	Filed Winding Self Inductance	H
L_q	Quadrature Axis Inductance	H
L_s	Self Inductance of the Stator	H
L_{ss}	Stator-Stator Mutual Inductance	H
M	Peak Stator-Rotor Mutual Inductance	H
N	Rotational Speed	rpm
P	Electric Power	W
p	Number of poles	-
P_{Lyp}	Solution of Lyapunov Equation	-
R_a, R_b, R_c	Resistance of the Stator Windings	Ω
R_f	Field Winding Resistance	Ω
R_o	Overall output impedance of the load and Battery	ω
T	Torque	Nm
T_e	Electromechanical Torque Produced by Motor	Nm
T_L	Load Torque	Nm
U	Vector of Model Inputs	-
u_n	Adaptive Control Signal	-
u_n	Base-Line Controller or Nominal Controller	-
V_a, V_b, V_c	AC Phase Voltages	V
V_{Bat}	Battery Voltage	V
V_d	Direct Axis Voltage	V
V_{dc}	Alternator DC Voltage	V
V_q	Quadrature Axis Voltage	V
K_P, K_I	Parameters of the Voltage regulator	V
V_{ref}	Alternator Reference Voltage	V
W_o	Unkown Parameters of the Fault or Uncertainty	-
$\tilde{W}$	Estimate of the Fault Parameters	-
X	State Vector	-
Y	Vector of Model Outputs	-
y	output	-

Abstract

On-Board Diagnostics provide vehicle drivers and repair facilities with information about the state of the various subsystems and enable self-diagnostic and reporting capability. Nowadays, the amount and type of diagnostic information available is limited to primitive or abstract diagnostic codes. With the transition towards electrification in powertrain, key components like alternators or traction motors still have their diagnostic information as abstract codes stored in the memory and can only be retrieved by maintenance service. Typically, these diagnostic measures are reviewed only in the event of severe fault situations and near breakdown.

Automotive electric machines represent the core of the driving and supply line of the vehicle; therefore, it is crucial to have an advanced and reliable condition monitoring system that facilitates on-going and direct fault detection and identification. This should be implemented during operation and before their total failure.

This work provides a novel multi-functional and model-based approach for fault detection and identification in automotive electric machines. This approach enables the vehicle to detect faults instantaneously long before the total breakdown of the components. Additionally, this approach utilizes the concept of digital twin to a great extent and parameter estimation as the core concept for fault detection. Followed by fault identification in terms of type and location, which is provided by machine learning classification algorithms.

This multi-functional approach extends its functionality to include fault tolerant control using model reference adaptive control to mitigate the effects of faults and recover the reference performance of the electric machine, thus ensuring a good and safe driving experience.

This work distinguishes itself from existing research by employing the available measurement in the vehicle, needless to implement additional sensors and providing deep analysis about the dynamic response of the machine. Secondly, incorporating the digital twin shifts the intensive computations to a high-power computational server. Third, this work represents one of the first attempts to implement fault-tolerant control to compensate the effects of faults within automotive electric machines themselves and during operation. Finally, the flexibility provided by this approach allows the transfer of information about the operating condition of the machine directly to the back-end diagnostic system which guarantees a comprehensive machine health assessment.

It is worth mentioning that the designed novel functions that interactively support each other, within the proposed approach, have provided successful results in detecting faults based on parameter deviations, identifying faults based on machine learning with very good accuracy exceeding 90 % of correct predictions, and powerful performance recovery provided by the adaptive control law.

Kurzfassung

On-Board-Diagnose liefert dem Fahrzeugfahrer und den Reparaturwerkstätten Informationen über den Zustand der verschiedenen Fahrzeugsubsysteme und ermöglicht Selbstdiagnose und Fehlermeldungen. Heutzutage sind Umfang und Art der verfügbaren Diagnoseinformationen auf einfache oder abstrakte Diagnosecodes beschränkt. Mit der Elektrifizierung des Antriebsstrangs werden die Diagnoseinformationen von Komponenten wie Lichtmaschinen oder Traktionsmotoren wichtiger und weiterhin als abstrakte Codes gespeichert. Diese Diagnoseinformationen können nur von den Diagnosewerkzeuge abgerufen werden. In der Regel werden diese Diagnosemaßnahmen nur bei schwerwiegenden Fehlern und Beinaheausfällen ausgewertet.

Elektrische Maschinen im Fahrzeug bilden das Kernelement des Antriebs und der Versorgung. Daher ist ein fortschrittliches und zuverlässiges Zustandsüberwachung System, das eine kontinuierliche und direkte Fehlererkennung und -identifizierung ermöglicht, unerlässlich. Dies sollte während des Betriebs aktiv sein und vor einem Totalausfall warnen.

Diese Arbeit bietet einen neuartigen multifunktionalen und modellbasierten Ansatz zur Fehlererkennung und -identifizierung in elektrischen Antriebe der Fahrzeuge. Dieser Ansatz ermöglicht es dem Fahrzeug, Fehler sofort zu erkennen, lange bevor die Komponenten vollständig ausfallen. Darüber hinaus nutzt dieser Ansatz in hohem Maße das Konzept des digitalen Zwillings und die Parameterschätzung als Kernkonzept der Fehlererkennung. Anschließend erfolgt die Fehleridentifizierung nach Typ und Ort, die durch Klassifizierungsalgorithmen des maschinellen Lernens bereitgestellt wird.

Dieser multifunktionale Ansatz erweitert seine Funktionalität um eine fehlertolerante Regelung mittels Modellreferenz adaptiver Regelung, um die Auswirkungen von Fehlern zu mildern und die Referenzperformance der elektrischen Maschine wiederherzustellen. Dadurch wird ein gutes und sicheres Fahrerlebnis gewährleistet.

Dieser Ansatz ist in einem dreistufigen, miteinander verbundenen Rahmen organisiert, der wie folgt lautet:

1. Modellbasierte Fehlererkennung
   - Diese Implementierung wird beispielhaft auf die Lichtmaschine angewendet, ist aber nicht darauf beschränkt. Diese Phase umfasst die folgenden Aspekten:
   - Herleitung eines äquivalenten und vereinfachten mathematischen Modells des Generators auf Basis einer dreiphasigen Synchronmaschine. Dabei wird die funktionale Ähnlichkeiten zwischen der dreiphasigen Synchronmaschine mit einem dreiphasigen Brückengleichrichter, und einem Gleichstromgenerator zur Herleitung des äquivalenten Modells genutzt.
   - Die Methode nutzt Parameterschätzungen als Grundlage für die Fehlererkennung. Das Hauptziel besteht darin, Folgendes zu generieren:
     - Ein parametrisiertes Modell, das die gesunde Maschine oder den Referenzbetriebszustand darstellt.
     - Ein parametrisiertes Modell, das die tatsächliche Maschine mit Fehlern oder Schäden darstellt.
   - Die Fehlererkennung erfolgt durch Vergleich der Parametersätze beider Modelle. Schäden können durch die Registrierung einer Abweichung der Parameterwerte erkannt werden, die einen vordefinierten Schwellenwert überschreitet.
   - Das mathematische Modell wird anhand von Messdaten parametrisiert. Diese Messungen werden von einem experimentellen Prüfstand generiert, der mit von Autor speziell entwickelten Versuchsreihen und bereitgestellten Komponenten betrieben wird.
   - Die Versuchsreihen werden sowohl mit gesunden Generatoren als auch mit Generatoren mit künstlich erzeugten Schäden oder Fehlern durchgeführt.
   - Die Parameter des mathematischen Modells werden anhand von Hardware basierten Messungen mithilfe eines Mehrzieloptimierung geschätzt, die die Methode der nichtlinearen kleinsten Quadrate verwendet und auf dem Trust-Region-Algorithmus basiert für die Suche des globalen Minimums.
   - Dieser Algorithmus sucht nach dem Parametersatz, der die zugewiesene Kostenfunktion minimiert.

2. Maschinelles Lernen zur Fehleridentifikation
   - Fehleridentifikation bezeichnet in diesem Zusammenhang die Bestimmung von Schadenart oder Fehlerart und -stelle.
   - Das Modell des maschinellen Lernen führt überwachtes Lernen durch und sagt auf dieser Grundlage den Betriebszustand des Generators voraus, ob er gesund ist, einen Schaden oder Fehler aufweist und welche Art von Schaden vorliegt.
   - Das Modell des maschinellen Lernens wird anhand gekennzeichneter Daten trainiert, die die Schadenart mit der Abweichung der mathematischen Parameter korrelieren.
   - Das Modell wird mithilfe mehrerer Klassifizierungsalgorithmen trainiert, um den besten Kandidatenalgorithmus hinsichtlich Genauigkeit und Rechenkomplexität zu ermitteln.
   - Die Klassifizierungsalgorithmen mit der höchsten Validierungs- und Testgenauigkeit sind Support Vector Machine und Wide Neural Networks.
3. Fehlertolerante Regelung auf Basis modellreferenzadaptiver Regelung
   - Die modellreferenzadaptive Regelung ist abgeleitet von der Lyapunov-Stabilitätsmethode. Der Hauptziel ist die Wiederherstellung beeinträchtigter Leistung und die Minderung der Auswirkungen von Fehlern oder Schäden.
   - Der vorgeschlagene Ansatz ist ein verallgemeinertes Konzept und wird zusätzlich zu den bestehenden Reglern in den Generatoren (Spannungsreglern) sowie in den Traktionsmotoren (feldorientierte Regelung) in Elektrofahrzeugen angewendet.
   - Bei diesem Ansatz wird der Tracking-Fehler zwischen dem Modell der gesunden Maschine (die in der Cloud läuft) und der tatsächlichen Maschine mit Fehler (im Fahrzeug) wird adaptiv minimiert. Indem die bestehenden Regler Parameter neu abgestimmt und fehlerbedingte Unsicherheiten minimiert werden.

Kernergebnisse

Der Hauptbeitrag dieser Dissertation besteht in der Bereitstellung eines neuartigen multifunktionalen und modellbasierten Ansatzes zur Fehlererkennung und -identifizierung in elektrischen Antrieben im Fahrzeug. Mit diesem Beitrag wurden die folgenden zentralen Forschungsfragen der Dissertation beantwortet:

- Wie lässt sich der aktuelle Stand der Fahrzeugdiagnose in ein fortschrittliches und zuverlässiges Zustandsüberwachungssystem umwandeln?
- Ist es möglich, die Referenzleistung der intakten Maschine trotz vorhandener Fehler wiederherzustellen?

Das Hauptziel dabei ist, kritischen Fahrsituationen zu vermeiden und ein gutes und sicheres Fahrerlebnis zu gewährleisten. Die wichtigsten Ergebnisse dieser Dissertation lauten wie folgt:

- Diese Dissertation bietet ein vereinfachtes und linearisiertes äquivalentes mathematisches Modell für die elektrische Maschine. Dieser Schritt dient der Vorbereitung der Fehlererkennungsmethode. Das äquivalente Modell umfasst Parameter, die eine aggregierte Darstellung mehrerer physikalischer Parameter liefern.
- Basierend auf dem Ersatzmodell ist es möglich, die Auswirkungen verschiedener Fehler im Electric Power Generation and Storage System des Fahrzeugs zu unterscheiden. Darüber hinaus hat dieses Ersatzmodell eine hohe Zuverlässigkeit und sehr gute Genauigkeit bei der Simulation der untersuchten Maschine entsprechend den ausgewählten Fahrprofilen bewiesen.
- Die Parameter des Ersatzmodells werden anhand experimenteller Daten und eines mehrzieloptimierung Problem geschätzt. Durch die Implementierung zusätzlicher Maßnahmen, ähnlich der Verwendung einer großen Anzahl von Anfangswerten, werden globale Minima und die Reproduzierbarkeit der Lösung gewährleistet. Auf dieser Grundlage beweist diese Methode ihre Wirksamkeit bei der Schätzung der Parameterwerte des Ersatzmodells sowohl für die gesunde als auch für die Maschine mit Fehlern oder Schäden.
- Die abschließende Analyse der Ergebnisse dieser Methode hat gezeigt, dass die Parameterwerte des Ersatzmodells stark vom Betriebspunkt der Maschine abhängen. Daher werden umfassende Lookup-Tabelle erstellt, die die vollständige Abbildung der Parameterwerte innerhalb des ausgewählten Betriebsbereichs des Generators enthalten.
- Eine umfassende Analyse der Wirkungsgradprofile Generatoren mit Schäden hat einen signifikanten Wirkungsgradverlust in einem beträchtlichen

Bereich von Ausgangsleistung und Drehzahl gezeigt. Dieser Wirkungsgradverlust je nach Fehlerart unterschiedlich ausgeprägt ist. Neben dem Wirkungsgrad wurden auch die Zeitbereichseigenschaften des Generators mit Schaden analysiert. Die Ergebnisse dieser Analyse zeigten eine signifikante Veränderung des dynamischen Verhaltens des Generators, unter anderem hinsichtlich Überschwingen, Einschwingzeit und stationärem Fehler.

- Mathematische Modelle werden für Generatoren mit unterschiedlichen Fehler- und Schadensarten parametrisiert. Die Ergebnisse der Analyse bestätigten die Auswirkungen von Fehlern auf die Parameterwerte des mathematischen Modells des Generators. Daher kann ein Fehler im Generator während des Normalbetriebs erkannt werden, indem Änderungen oder Abweichungen der Parameterwerte im Vergleich zu den Werten des mathematischen Modells der intakten Maschine registriert werden.
- Die Analyse der Parameterwerte der verschiedenen Generatoren hat gezeigt, dass jede Fehlerart ein individuelles Muster bei der Beeinflussung der Parameterwerte aufweist. Darüber hinaus variiert dieses Muster je nach aktuellem Betriebspunkt sowie in Bezug auf den einzelnen Parameter selbst. Darüber hinaus führen Fehler nicht zwangsläufig zu einer Erhöhung der Parameterwerte, sondern es wurde auch eine Verringerung festgestellt.
- Weitere Untersuchungen haben gezeigt, dass der Schwellenwert, ab dem eine Parameterabweichung als fehlerbedingt angesehen wird, bei 2% liegt. Dieser Schwellenwert ist auch notwendig, um Toleranzen auszuschließen, die durch die Lösung des Optimierungsproblems verursacht werden.
- In dieser Arbeit werden Klassifizierungsalgorithmen für maschinelles Lernen implementiert, um Fehler hinsichtlich Art und Ort zu identifizieren. Der Grund für diese Wahl ist ihre effektive Fähigkeit, die verborgenen und komplexen Muster zwischen Parameterabweichungen und der Art des Fehlers zu verstehen.
- Es wurde eine vergleichende Studie zur Leistung verschiedener Klassifizierungsalgorithmen bei der Vorhersage des Betriebszustands des Generators durchgeführt. Diese Studie hat gezeigt, dass die auf Support Vector Machine und breiten neuronalen Netzen basierenden Klassifizierungsalgorithmen die höchste Vorhersagegenauigkeit von über 90% erreichten. Diese hohe Genauigkeit belegt die erfolgreiche Implementierung des maschinellen Lernens zur Fehleridentifizierung.

- Diese Arbeit erweitert die Funktionalität des neuartigen Ansatzes um die fehlertolerante Regelung, die als dritte und letzte Stufe des Rahmens gilt. Dies ist einer der Hauptbeiträge dieser Dissertation. Darüber hinaus ist die Implementierung auf Basis der modellbasierten adaptiven Regelung ist einer der ersten Versuche im Bereich elektrischer Antrieben, der in erster Linie darauf abzielt, die Referenzleistung wiederherzustellen und die Auswirkungen von Fehlern oder Schäden innerhalb des Antriebs während des Betriebs zu mildern.
- Das adaptive Regelgesetz hat sich als erfolgreiche Ergänzung sowohl zum PID-Spannungsregler im Generator als auch zur feldorientierten Regelung im Traktionsmotor erwiesen.
- Die erfolgreiche Leistung von diesem Adaptiven Regler als Methode zur Kompensation und Milderung der Auswirkungen von Fehlern und Schäden wurde anhand umfassender Testfälle sowohl für den Generator als auch für den Traktionsmotor nachgewiesen.
- Einer der Hauptvorteile von Modellreferenz-adaptiver Regler besteht darin, dass es auf bestehenden Controllern implementiert wird, ohne dass Struktur und Design der vorhandenen Controller geändert werden müssen.

Diese Arbeit unterscheidet sich von bestehenden Forschungsarbeiten dadurch, dass sie die schon verfügbaren Messdaten im Fahrzeug nutzt, ohne zusätzliche Sensoren, und eine umfassende Analyse des dynamischen Verhaltens der Maschine ermöglicht. Zweitens verlagert die Integration des digitalen Zwillings die rechenintensiven Berechnungen auf einen Hochleistungsserver. Drittens stellt diese Arbeit einen der ersten Versuche dar, eine fehlertolerante Regelung zu implementieren, um Fehler in elektrischen Fahrzeugmaschinen selbst und während des Betriebs zu kompensieren. Schließlich ermöglicht die Flexibilität dieses Ansatzes die Übertragung von Informationen über den Betriebszustand der Maschine direkt an das Back-End-Diagnosesystem, was eine umfassende Bewertung des Maschinenzustands gewährleistet.

Zusammenfassend lässt sich sagen, dass diese multifunktionale, modellbasierte Fehlererkennung und -identifizierung neben der fehlertoleranten Regelung für elektrische Maschinen im Automobil einen wichtigen Beitrag zur Funktionalität der Fahrzeugdiagnose leistet. Sie gilt als erster Schritt zur Vorbereitung auf die Anforderungen zukünftiger Fahrzeuge.

1 Introduction

Faults in automotive electric machines such as the alternator and the traction motor have a direct effect on the performance, energy supply, driver comfort, as well as the functional safety. Therefore, it is crucial to have an advanced and reliable condition monitoring system that facilitates continuous and direct fault detection and identification in these machines, which is implemented during operation and before their total failure. For these machines as they represent the core of the driving and supply line of the vehicle any downtime, interruption or performance degradation will have a direct impact on the driving experience as well as the performance of the entire vehicle.

Identifying and understanding the nature and source of faults have pivotal importance in developing techniques and measures to mitigate their effects and compensate for degraded performance of the machine and thereby providing a great assessment for predictive maintenance and repair operations.

Digital transformation in the automobile industry have paved the way towards introducing the emerging concept of Digital Twin (DTw). This concept takes a great advantage of the already existing Internet of Things (IoT) - sensors and actuators within the vehicle in addition to the powerful data analysis and visualization tools. Moreover, DTw is regarded as key enabler of the Software-Defined Vehicle (SDV), which allows continuous connectivity to the cloud, automation and updates of features and services.

This research work utilizes the concept of DTw within a novel multi-functional approach for fault diagnosis and recovery particularly, designed for automotive electric machines. This work leverages the predictive maintenance measures and implementation for these machines.

The implementations and investigations in this research work contribute in the front loading towards realizing the cutting-edge technologies and concepts of vehicle self diagnostics, DTw and SDV.

A. Salah, *Model-Based Fault Diagnosis and Fault-Tolerant Control*,
Wissenschaftliche Reihe Fahrzeugtechnik Universität Stuttgart,
https://doi.org/10.1007/978-3-658-50108-2_1

1.1 Motivation and Objective

Vehicle's self diagnostic and self reporting capability which is known as On-Board Diagnostic (OBD), gives the vehicle owner and repair facilities information about the status of the various vehicle sub-systems. However, the amount and type of diagnostic information available via OBD are limited to primitive or abstract Diagnostic Trouble Codes (DTC). In general, OBD provides information on the status of the engine and emission system and triggers a malfunction indicator light in case of faults. On the other hand, Off-Board Diagnostic (UDS) handle the diagnostics related to every other Electronics Control Unit (ECU) in terms of fault codes that are stored in the memory. For this type, instant reporting is not carried out and fault codes can be retrieved by service maintenance. Consequently, the diagnostic measures provided by both protocols are signalized and analyzed only at severe fault situations or near breakdown.

This work is motivated by the current state of diagnostics in the automotive domain. As it introduces a novel multi-functional approach for Fault Detection and Identification (FDI). This approach enables the vehicle to detect faults within the automotive electric machines instantaneously, and long before the total breakdown of the components. In addition to fault detection, fault identification in terms of type and location takes place with the help of the cutting-edge technology of Machine Learning (ML). The flexibility of this approach allows the transfer of information about the operating condition of the electric machine directly to the back-end.

This multi-functional approach extends its functionality to include Fault-Tolerant Control (FTC) in order to mitigate effect of faults and recover the reference performance of the electric machine. This approach incorporates Digital Twin (DTw) and shifts intensive computations to a high-power computational server to provide a comprehensive machine health assessment.

This research work contributes toward realizing and preparing for the future diagnostic systems that adapt to the future automotive functions such as automated driving vehicles.

1.2 Scope

This research work considers automotive electric machines such as the alternators and traction motors and aims to provide a generalized fault detection, identification, and performance recovery techniques. The faults that are considered here occur very often in the aforementioned machines during operation. In spite of the fact that these faults do not include the total failure of the component, however, their progress with time will lead eventually to the total failure or breakdown of the component and consequently the electric machine. At the same time, their presence causes an observable degradation in the provided performance.

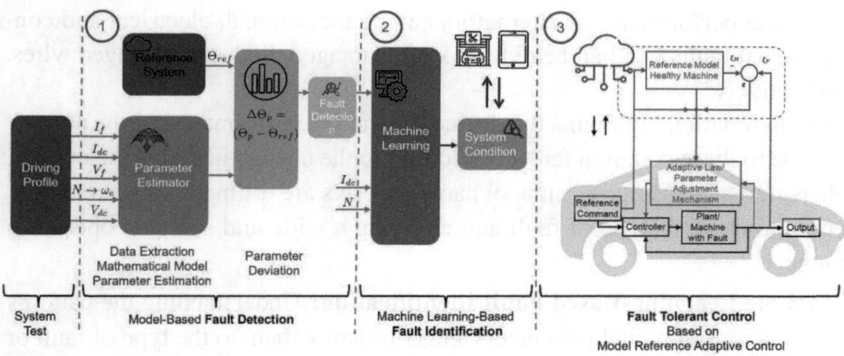

Figure 1.1: General Framework for Fault Detection, Identification and Fault Tolerant Control of Automotive Electric Machines

This dissertation presents a novel approach in the field of automotive electric machines for fault detection and identification to fault-tolerant control. The framework of this approach is shown in Figure 1.1. The three interconnected core topics and key concepts of this framework are described in the following.

- **Model-Based Fault Detection:** This implementation is applied to the alternator as an example but is not limited to. The technique employs parameter estimation as a base for detecting faults. The main goal here is to generate a parametrized model which represents the healthy machine or the reference operating state in addition to a model which represents the actual machine-

with-faults. By comparing parameter sets of both models, faults can be detected by registering a deviation in parameter values which exceeds a predefined threshold. It is worth mentioning that the mathematical model, which represents a simplified and equivalent version of the physics-based one, is parameterized utilizing measurement data. These measurements are obtained from an experimental test-bench which is operated using specially designed components and series of experiments. In addition to that, the experiments are repeated on alternators with artificially applied faults. Consequently, parameters of the mathematical model are estimated based on a multi-objective optimization problem which employs least squares algorithms. As anticipated, these algorithms search for the parameter set which minimizes the assigned cost function. In this work a comprehensive study is presented about the performance of alternators having mechanical, electrical and control signal faults like ball bearing-flaking, damaged diodes to damaged wires, respectively.
This dissertation highlights the degradation in the performance of the alternator due to these common faults in addition to the change in their time domain characteristics. Moreover, model parameter sets are estimated for alternators containing every type of fault and covering a wide and selected operating range.

- **Machine Learning-Based Fault Identification:** Understanding the patterns of deviations of model parameters and correlating them to the type of fault or in other words performing fault identification is not trivial, a good candidate for this role is a machine learning classification algorithm. The former is capable of understanding the hidden and complicated patterns between several observations and, accordingly, capable of identifying various faults based on parameters variation. In this work, a supervised machine learning model is trained utilizing labeled data correlating the type of fault to the deviation in parameters. The classifier model then provides predictions about the current operating condition of the actual alternator incorporating new data and identifying the type of faults if they exist. This implementation of machine learning for fault identification has a high prediction high accuracy exceeding 90 %
- **Fault Tolerant Control Based on Model Reference Adaptive Control:** Model Reference Adaptive Control (MRAC) is implemented in this disserta-

tion to handle the problem of Fault-Tolerant Control (FTC) in automotive electric machines. It aims primarily to recover the degraded performance and mitigate the effect of faults. The proposed approach is a generalized concept and is applied on top of the existing controllers in the alternators (voltage regulators) as well as in the traction motors (field-oriented control).
In this work the analogies between model uncertainties and parameter deviations caused by faults are utilized to further develop the implementation of MRAC actively in the field of FTC. In this context, the reference model represents the healthy operating condition of the machine that is required to be maintained by the machine-with-faults. This implementation effectively demonstrates the deployment of the digital twin concept and and its integration with cloud connectivity.
Within this approach the tracking error between the healthy model of the machine and the actual fault-containing one, in the vehicle, is minimized adaptively by re-tuning the existing controller parameters and mitigating uncertainties caused by faults. This concept has demonstrated successful results and performance in the domain of fault-tolerant control for both alternators and traction motors.

This research work is limited to utilizing the main concept of a digital twin and is not intended to address its technological aspect, such as:

- Connection mechanisms to the cloud and its specification.
- Aspects regarding data transfer-rate and Over-the-Air-Update (OTA-Update).

In addition to that, the implementation of ML is limited to utilizing the already available algorithms. The development of new concepts and algorithms are beyond the scope of this dissertation.

Finally, this contribution aims to give a valuable insight into the performance degradation of the electric machine due to presence of faults and a better understanding of the resulting dynamic behavior and based on that a unique FTC concept is derived.

1.3 Literature Review and State of the Art

Fault detection and identification for automotive electric machines are gaining more attention. Whether these are the traction motors or the alternators within the electric power generation units, a faultless machine which delivers the expected performance, ensures the functional safety and comfort of passengers in addition to the desired efficiency. Therefore, it is crucial to develop a robust Fault Detection and Identification (FDI) technique based on the available measurements in the vehicle and without introducing additional complexity. These measurements reflect the current condition of the machine, allow data analysis and comparison to reference systems or operating conditions.

Following FDI, is to design Fault-Tolerant Control (FTC) control scheme in order to mitigate the effect of faults and compensate for the impaired performance.

This literature review is divided into three main parts concerning the topics; fault detection, identification and fault tolerant control for electric machines. Then followed by a comparative evaluation to highlight the excel and advantages of the novel approach introduced within this dissertation.

Signal-Based Fault Detection and Identification

A great amount of research work is accomplished based on analyzing the signals and their frequency spectrum of various measurements within the electric machine. By analyzing these signals, anomalies, deviations and fault patterns can be identified providing a good insight into the operating state of the machine. Additionally, these signals encompass the measurements of currents, voltages to vibrations and temperature.

One of the most common and widely implemented technique is fault detection based on analyzing the current signature. This technique is applied by the research in [1] to detect the stator-turn faults in an automotive claw-pole alternator. The former is achieved by designing a fault detecter that analyzes only the DC output current spectrum and observing the harmonics. It is concluded that, the amplitude of the fault-current harmonic is related to the fault level. This technique however, requires a proper definition of the threshold value which may vary with respect to loads. Additionally, this technique is implemented within an environment having an optimized signal to noise ratio therefore, there is a need to evaluate this technique within an experimental environment similar to the vehicle. Similar to the previous research, analyzing the spectrum of the residual stator current is developed by [2] to detect bearing faults in high speed synchronous machines. Moreover, this work generates the current spectrum of the healthy machine for comparison purposes. Although this work shows successful results, however, the isolation and identification of bearing fault based on the stator current, in the presence of other faults such as inter-turn short and open circuits should be evaluated.

The method described in [3] analyzes components of the stator current spectra or both healthy and faulty machines in order to detect inter-turn short circuit in induction machine. The method shows that this fault causes a rise in some of the frequency components of the line current. A frequency pattern is extracted directly from the spectrum of the magnetic flux density and is correlated to the short-circuit fault diagnosis in [4]. The occurrence and the number of short circuited turns are predicted using support-vector machine as a classifier. On the other hand, motor starting current is observed in [5] and analyzed in time-frequency domain to detect features that are related to faults however specific types and locations of faults are not analyzed.

Alternatively, rotor current is analyzed to detect diode fault in [6]. The method develops a current restructuring approach where the magnitude ratio of the second harmonic to the fundamental is obtained by band-pass filters and utilized to detect the fault.

Additionally, the signature of the electromagnetic field that passes through stator tooth, is utilized for detecting inter-winding faults in Permanent Magnet Synchronous Machine (PMSM) and is developed by [7]. The magnetic field is captured by a search coil and a mathematical model.

The Park transformation is an additional primary technique that falls under the current signature category. It is commonly implemented to detect faults in electric machines based on rotating reference frames of AC phase currents such that they become DC signals. Subsequently, the modified currents are examined to identify specific features associated with faults. These techniques demonstrate notable accuracy in identifying different faults; however, from the vehicle's perspective, it's important to note that measurements of phase currents are not accessible for the diagnostic system. In [8], the detection of inter-turn short circuit fault in permanent magnet-assisted synchronous reluctance machine is based on the transformed current residuals. Then a comparison to the current signature of the healthy machine is obtained. Based on that, the fault indicator is determined by the magnitude of the negative-sequence dc components of the I_q and I_d current residuals. The key benefits of this method is that the signature of the fault current is enhanced and the impact of the model errors, other harmonics, and the speed transients on turn-fault detection are minimized.

Parks vector transform is also implemented by [9] to diagnose the rotor condition in a squirrel cage induction motor. This approach suggests a potential correlation between power supply asymmetry and a damaged rotor. Similarly, the study conducted by [10] has investigated the same fault of damaged rotor in addition to stator short-circuit within induction machines. The proposed method transforms the supply currents into the stationary reference frame.

In addition to current signatures, voltage signatures are also used to detect faults. The signature of broken bars on the supply voltage of induction motors is investigated by [11]. The research employs additive models to quantify the influence of faults.

High resistance connection in PMSM is explored by [12]. The study proposes a method based on correlation with the zero sequence voltage component, in addition to the stator currents and the resistance of the faulty phases. Fault detection is described by solving a binary linear equation. This study, despite its effectiveness in detecting the mentioned faults, encompasses relatively high computational complexity.

An initial simulation study is offered by [13] that examines the magnitude of the ripple in the output voltage to the presence of faults in the alternators. Although the method does not associate the magnitude to a specific fault type, it provides a Pareto analysis to scale the severity of the fault based on the shape and magnitude of the ripple.

Fault diagnosis based on wavelet transformation is another widely implemented method. A detailed survey on the implementation of wavelet in induction motor fault diagnosis is provided by [14]. However, this method has challenges with respect to implementation in real-time systems, such as memory usage, and involves multiple operations, which can be costly in terms of time and space. In addition, some wavelet techniques require access to the complete signal before analysis can be performed. As a result this technique despite its efficacy is less suitable for implementation during the operation of the vehicle.

Vibration signal analysis is another widely used signal-based method for fault detection in electric machines based on vibration signal analysis. The main concept of this technique is to analyze the acceleration signals from the main two axes. Subsequently, the envelope spectrum analysis and spectral kurtosis are employed for the diagnosis of bearing faults as detailed in [15, 16, 17]. This method encompasses computing four critical frequencies that are related to the position of fault on the bearing. This research has been extended in [15] to include a machine learning classification algorithm. On the other hand, the research in [18], handles the problem of rotational speed fluctuations in the crankshaft of an engine and provides a prototype to diagnose the operating frequency of the generator. The developed method correlates the output electrical frequency with the spectrum of the peak envelope. After that, a pattern recognition algorithm has been applied.

Many researches have combined the signal-based method with several Artificial Intelligence (AI) techniques. This is a reasonable addition as these techniques assess the classification and recognition process. For example, the research in [19] and [20] employs fuzzy logic in combination with the Park transformation to detect turn-to-turn short circuit in PMSM. Neural networks are applied in [21] to monitor and learn the degradation profile of components in an induction motor. This approach enables the identification of faults such as insulation degradation and broken rotor bars. Finally, gearbox faults are detected by analyzing vibration signals in combination with a machine learning classifier.

The above described signal-based methods, despite their efficacy in detecting a wide range of faults, have the following drawbacks from the perspective of implementation within the vehicle:

- The need for additional sensors or measurement configurations in order to analyze variables that are not available in the main diagnosis and communication of the vehicle. Implementing these configurations and sensors will lead to a considerable increase in manufacturing and maintenance costs.
- The additionally required sensors need to possess relatively high resolution and stable data transfer medium, these issues are limited within the vehicle.
- The capability of current communication in vehicles using CAN Bus and LIN Bus for actuators provides a limited transfer rate.
- Analysis in the frequency domain and the spectra of the measured signals require larger memories and powerful processors, and consequently increased cost.
- The results of these methods require a deep analytical investigation in the frequency domain and spectral envelope, and the application of an automated algorithm is still questionable.
- Real-time implementation needs to be considered, some of the analysis techniques provide results based on the complete course of signal or phrased differently are more suitable for offline analysis.

Taken together, all the above mentioned methods of signal-based FDI are a custom-design, which means that they are based on individual machines and faults and are dedicated to specific cases. An efficient signal-based FDI for automotive electric machine needs to consider all the above requirements.

Model-Based Fault Detection and Identification

This concept is based mainly on the involvement of a mathematical model of the system or machine under consideration and therefore all the derived techniques are related to this model. The essence of this method utilizes a comparison between the actual system, actuator, or even sensor, then a simulation model derived based on several concepts. In addition to that, based on measured input signals and output signals, detection methods generate residuals, parameter estimates, or state estimates, and these are called features [22]. Moreover, a continuous comparison between measured states, inputs, and outputs is carried out in order to analytically register symptoms within the above-described features that are related to faults.

Model-based fault detection involves methods similar to residual generators, including parity equations, Kalman filter, and parameter estimation to state observers. Fault modeling and system identification are also essential parts of this method [23].

A method for fault diagnosis in automotive Electric Power Generation and Storage (EPGS) system is developed by [24] and [25] to detect and identify faults in the alternator. Both studies utilize parity equations to provide a comparison between the reference model and an actual alternator with fault. Furthermore, in [25] detailed and simplified modeling of the EPGS system is provided alongside parameter estimation. Symptoms for identifying faults similar to voltage regulator, transmission belt, and rectifier are analyzed. This research is further developed to approach the diagnostic problem with a hierarchical method, thus providing a supervisory diagnostic system that is more suitable for implementation in the vehicle. It is interesting to investigate, however, the presence of more than one fault using these parity equations.

An extended Kalman filter is employed by [26] to detect anomalies in induction machines. The research proposed an extended model that includes additional states and disturbances that enable the detection of discrepancies between the model and the real system. The model introduced in [27] also utilizes the Kalman filter for online estimation of friction in a dc motor. Subsequently, the deviation from the nominal system is used to indicate faults. Kalman filter techniques are also utilized in [28] to estimate the rotor resistance of an induction motor based on the measured stator currents and voltages. Additionally, the broken bar fault is detected by comparison with the observer representing the healthy machine. A similar approach is implemented by [29] but for the detection of bearing faults. By Implementing Kalman filter and observers despite their effectiveness in estimating states, one should consider the computational complexity and their sensitivity to noise and modeling inaccuracies.

Residuals-based method is implemented by [30] for EPGS to detect belt-slip and estimate its size. The study suggests that there is a relationship between the duty cycle and the voltage drop at the output. Based on this, the fault symptoms are collected by registering the deviation from the nominal line. This method encompasses several approximations and might not be generalized.

Equivalent model and parameter estimation is introduced by [31] to detect induction motor faults. For parameterizing the equivalent multi-coupled circuit, feedforward neural networks are used. This approach has the advantage of implementing simplified reference models and utilizing the capabilities of Neural Networks (NN) to estimate the parameters.

There is no doubt that signal-based methods for FDI are more commonly implemented and investigated. This is evident from the number of cited research studies implementing this type with respect to the model-based ones. On the other hand, the signal-based techniques, despite their efficacy, suffer a lot of drawbacks from the perspective of vehicle operation. At the same time, the model-based techniques offer considerably more flexibility, simplification, and the ability to adapt their development to the modern requirements of the future vehicle. However, this domain still needs to be thoroughly explored.

Fault Tolerant Control FTC for Electric Machines

A reconfigurable FTC scheme is designed in [32] for a motor drive to manage voltage sensor failure. The research work suggests that in the case of voltage sensor failure the drive transfers the control smoothly to the inner-control loop represented by field-oriented control. In [33] an observer-based FTC is presented to manage the failure of the current sensor in the induction motor controlled based on a double torque control method.

The research conducted by [34] presents a switching-based FTC for an induction motor. The FTC controller comprises two switching modes; one for nominal control under normal operating conditions and the second is a robust control to handle faulty conditions. In addition to the robust controller, a speed estimator and an open-loop flux observer are designed to compensate sensor faults. A flux observer and speed estimator are also developed by [35] based on the approximation of the magnetic characteristic slope of an induction motor to the mutual inductance value, ensuring a senseless operation. On the other hand, correction and reconstruction of the hall sensor signal in a DC motor are suggested by [36].

The work in [37] addresses the problem of an open phase in the supply of an induction motor and proposes a FTC that encompasses a new current transformation scheme. This proposed transformation converts the asymmetrical equations caused by faults into symmetrical ones. Comprehensive literature reviews and surveys about FTC and its applications are provided by [38, 39, 40] and [41].

In the domain of FTC for electric machines, the majority of work targets compensation of sensor faults. This aspect is crucial in the operation of an electric machine. However, from another point of view, faults inside electric machines are rarely investigated specifically in the automotive domain, implying a gap that needs to be explored. The novelty of the work presented in this dissertation arises from addressing this issue.

Model Reference Adaptive Control (MRAC) for Electric Machines

Most implementations of MRAC are to increase the accuracy and improve the performance of the electric machine with the existing controller and not necessarily to handle uncertainties and faults. The control concept MRAC has been identified by [42] to be a good candidate for sensor-less motor operations in detecting the position and speed of PMSM. Similarly, the work presented in [43] for sensor-less speed control of asynchronous machine. An observer-based MRAC is implemented in [44] in combination with fuzzy logic to accurately determine the position of the rotor of a PMSM. On the other hand, the research in [45] integrates MRAC with sliding mode control to estimate the induction motor speed. MRAC is also proposed for the speed control of the double fed induction machine according to [46] and PMSM according to [47]. Finally, MRAC is also used as a compensation method to eliminate steady-state position error for PMSM. The detailed derivation and the proof of the stability of MRAC are provided by [48] and [49].

1.4 Dissertation Novelty

This dissertation presents a novel multi-functional approach comprising a model-based fault detection, machine learning-based fault identification and fault tolerant control specifically for automotive electric machines. The model-based FDI method employs parameter estimation as a base for detecting faults. The main goal here is to generate a parametrized model which represents the healthy machine or the reference operating state in addition to a model which represents the actual machine-with-fault. By comparing parameter sets of both models, faults can be detected by registering a deviation in parameter values, which exceeds, a pre-defined threshold. The mathematical model, which represents a simplified and equivalent version of the physics-based one, is parameterized utilizing measurement data and based on a multi-objective optimization problem. It is worth mentioning, that computations of the optimization problem are shifted to the cloud and carried out on a regular basis and not continuously, thus reducing the computational cost significantly. This method is distinguished from the previously mentioned research work by utilizing the existing measurement signals within the vehicle, needless to implement additional sensors, the complex computations are accomplished by a high-power computational server that is assigned to the vehicle. Additionally, the method provides a complete mapping of the electric machine over the full operating range deploying the digital twin concept, which is one of the most emerging technologies in the automotive industry. Finally, this method is a generalized method as it detects any type of faults which results in parameter deviation. Since it covers a wide spectrum of components within the EPGS system of the vehicle.

This multi-functional approach incorporates a mechanism to identify those faults in terms of type and location. This mechanism is based on ML classification model which performs supervised learning. This effective method identifies patterns in the model parameter deviations and associates them with faults. This implementation of ML for fault identification has a high accuracy exceeding 90 % of correct predictions.

The work in this dissertation extends its scope to include fault tolerant control that is based on MRAC. This adaptive control law enables the electric machine to recover the desired performance within the defined limits ensuring an adequate driving experience. Moreover, the technique significantly benefits from the established reference model that operates parallel to the hardware. This method also adaptively minimizes a tracking error with respect to the actual machine with fault. It is worth mentioning that this method is not limited to a certain machine or fault, but rather it is generalized, and its successful performance has been demonstrated on the alternator, as well, as on the traction motor and against mechanical as well as electrical faults.

To the best of the author's knowledge this is one of the first implementations of FTC based on MRAC specifically for automotive electric machines focusing on the actuator not the sensors. In addition to that, this novel multi-functional approach of model-based fault detection, machine learning-based fault identification and fault-tolerant control is an outstanding technique as it provides a comprehensive fault diagnosis scheme, extends beyond to recover the reference performance and adapts along with the emerging technologies of software-defined vehicle and digital twin.

1.5 Dissertation Outline

This dissertation comprises six chapters which are described as follows:

Chapter 1 presents the main topic of this dissertation with projection on the current technological development in the automotive domain. The main objective and scope are introduced. In addition to detailed literature review highlighting the current state of the art and the novelty of this research work.

Chapter 2 derives the mathematical model of Electric Power Generation and Storage (EPGS) in detail and in simplified form. Then the experimental setup is introduced, and finally, the parameter estimation problem is presented.

Chapter 3 introduces the model-based fault detection method, then the most common faults in automotive electric machines are highlighted followed by a detailed analysis on the effect of faults on the performance of the alternator and the values of the parameters of the mathematical model.

Chapter 4 presents the machine learning-based fault identification method where the model and its inputs and outputs are introduced. Then a detailed comparative study is presented on the performance, validation, and prediction results of several ML models. Followed by a comprehensive assessment of the selection of features and their sensitivity in the model. In addition, several recommendations are presented to increase the accuracy of the identification model.

Chapter 5 Model Reference Adaptive Control (MRAC) as a method to recover the healthy behavior of the electric machine and mitigate the influence of faults. This chapter includes a detailed derivation of the mathematical model of MRAC combined with two main controllers; the PID voltage regulator of the alternator and the Field Oriented Control (FOC) of the traction motor. The effectiveness and successful implementation of the adaptive control law are demonstrated with the help of four case studies and simulation results.

Chapter 6 provides a summary and overview of the entire dissertation, in addition to the conclusions and the outlook for future research work.

2 Automotive Electric Power Generation and Storage System

The Electric Power Generation and Storage (EPGS) system serves as an essential component in vehicles, primarily comprising a power storage unit (battery), a power generation unit (alternator), and various electrical consumers devices that can collectively be represented as total electric load. Figure 2.1 illustrates a schematic diagram of EPGS. This diagram demonstrates that the rotational speed is delivered by Internal Combustion Engine (ICE), while the alternator generates electricity to supply the various loads and electronics.

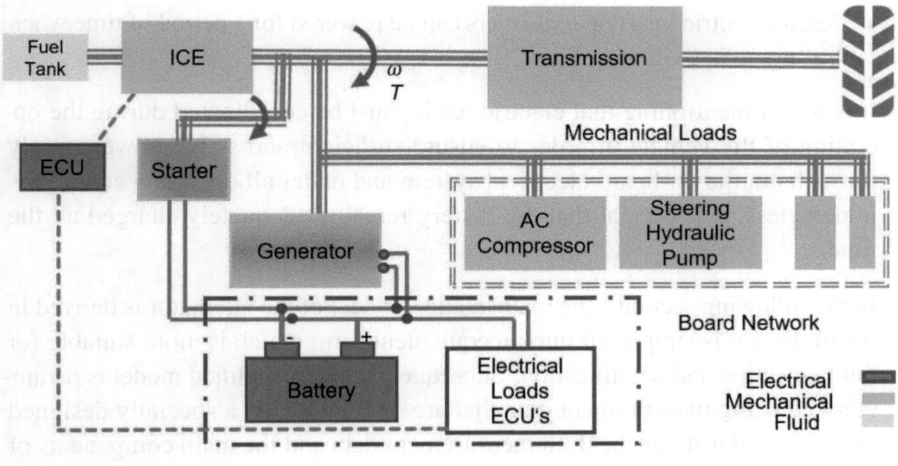

Figure 2.1: Electric Power Generation and Storage (EPGS)

In this system, the battery provides electric power for the starter motor of the Internal Combustion Engine (ICE), covers the shortage in the electric power supply and protects sensitive electrical and electronic components from voltage peaks and fluctuations. In addition to that, the battery satisfies transient power requirements.

A. Salah, *Model-Based Fault Diagnosis and Fault-Tolerant Control*,
Wissenschaftliche Reihe Fahrzeugtechnik Universität Stuttgart,
https://doi.org/10.1007/978-3-658-50108-2_2

The alternator is typically a 3-phase synchronous machine, equipped with a full bridge rectifier to provide DC electric power (I_{dc}, V_{dc}) for both the battery and the cumulative load in the vehicle. The alternator is required to supply electrical power for all types of electric loads and according to their demand with respect to time, whether it is continuous, intermittent, short- or long-term.

The supplied loads or units vary from heavy to light loads, and sensitive loads for example; the lighting, ventilators, air-conditioning unit, heaters, and many others. In general, the generator output, battery capacity, starter motor and the electrical consumers need to be carefully matched in such a way that:

- A balanced electrical system is created.
- The engine can be reliably started.
- Certain electric loads or consumers can be powered for a period of time when the engine is switched off.

It is worth mentioning that electric loads must be coordinated during the operation of the vehicle in order to ensure sufficient and stable power supply through out the on-board electrical system and under all operating conditions nonetheless, it is crucial that the battery remains adequately charged all the time.

In the following sections, the mathematical model of the alternator is derived in detail, then it is simplified into an equivalent form which is more suitable for fault detection and identification. Subsequently, this simplified model is parameterized using measurement data that are obtained from a specially designed experimental test bench. Both measurement data and the main components of the test bench are presented in Section 2.2. Finally, the performance and results of the alternator simulation model are presented and evaluated at the end of this chapter in Section 2.3.

2.1 Mathematical Model of Alternator

The alternator in modern vehicles is typically a claw-pole 3-phase synchronous machine. In this section, the mathematical model of the automotive alternator is derived in detail, and then in the following section it is simplified into an equivalent form that is more suitable for fault detection and diagnosis.

The alternator can be represented by an equivalent circuit diagram that is shown in Figure 2.2. The alternator is required to supply electrical power for all types of electric loads and according to their demand with respect to time.

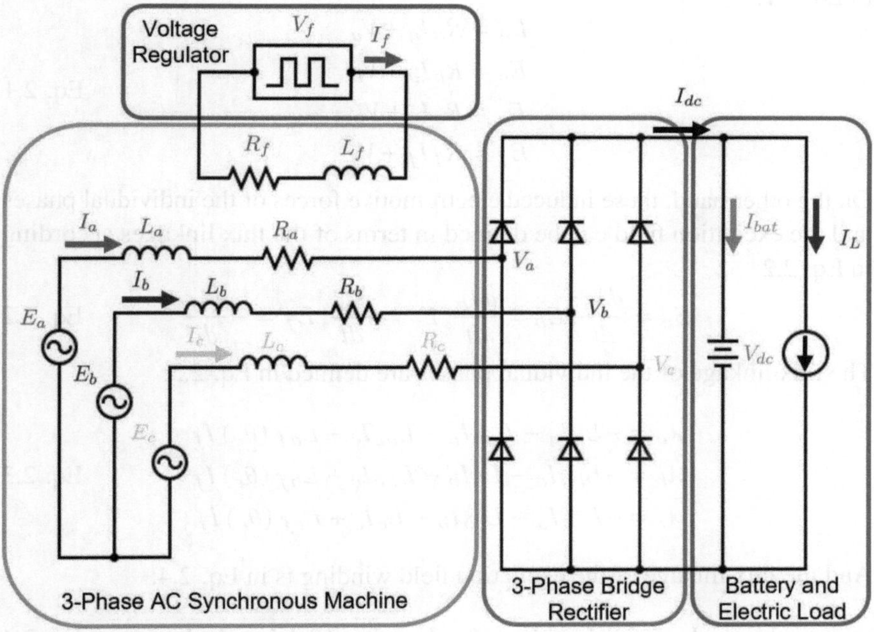

Figure 2.2: Simplified Circuit Diagram of Alternator and Bridge Rectifier

The following subsystems are included in this equivalent circuit diagram:

- Voltage regulator for the excitation field.
- Three-phase AC synchronous machines.
- Three-phase rectifier bridge.

The following mathematical derivations are based on the study provided by [50] and [25].

2.1.1 Detailed Mathematical Model of the Alternator

3-Phase AC Synchronous Generator

The induced electromotive force of the individual phases is derived based on the coupled circuit between the stator and the rotor as in [50] and according to Eq. 2.1:

$$\begin{aligned} E_a &= R_a I_a + V_a \\ E_b &= R_b I_b + V_b \\ E_c &= R_c I_c + V_c \\ E_f &= R_f I_f + V_f \end{aligned} \qquad \text{Eq. 2.1}$$

On the other hand, these induced electromotive forces of the individual phases and the excitation field can be defined in terms of the flux linkages according to Eq. 2.2:

$$E_a = \frac{d\lambda_a}{dt}, E_b = \frac{d\lambda_b}{dt}, E_c = \frac{d\lambda_c}{dt}, E_f = -\frac{d\lambda_f}{dt} \qquad \text{Eq. 2.2}$$

The flux linkage of the individual phases are defined in Eq. 2.3:

$$\begin{aligned} \lambda_a &= -L_a I_a - L_{ab} I_b - L_{ac} I_c + L_{af}(\theta_e) I_f \\ \lambda_b &= -L_{ba} I_a - L_b I_b - L_{bc} I_c + L_{bf}(\theta_e) I_f \\ \lambda_c &= -L_{ca} I_a - L_{cb} I_b - L_c I_c + L_{cf}(\theta_e) I_f \end{aligned} \qquad \text{Eq. 2.3}$$

And the flux linkage of the excitation field winding is in Eq. 2.4:

$$\lambda_f = -L_{fa}(\theta_e) I_a - L_{fb}(\theta_e) I_b - L_{fc}(\theta_e) I_c + L_f I_f \qquad \text{Eq. 2.4}$$

The stator-rotor mutual inductances of the machine are described according to Eq. 2.5:

$$\begin{aligned} \lambda_{af}(\theta_e) &= L_{fa}(\theta_e) = M\cos(\theta_e) \\ \lambda_{bf}(\theta_e) &= L_{fb}(\theta_e) = M\cos(\theta_e + \Phi) \\ \lambda_{cf}(\theta_e) &= L_{fc}(\theta_e) = M\cos(\theta_e - \Phi) \end{aligned} \qquad \text{Eq. 2.5}$$

The phase angle θ_e of the alternator is defined in terms of the mechanical angular displacement θ_m and the number of poles p in Eq. 2.6:

$$\theta_e = \frac{p}{2}\theta_m \qquad \text{Eq. 2.6}$$

Re-writing Eq. 2.1 in matrix notation, the induced electromotive force of the individual phases is described by Eq. 2.7:

$$\mathbf{E} = \mathbf{RI} + \mathbf{V} \qquad \text{Eq. 2.7}$$

Where,

$$\mathbf{E} = \begin{bmatrix} E_a \\ E_b \\ E_c \end{bmatrix}, \mathbf{V} = \begin{bmatrix} V_a \\ V_b \\ V_c \end{bmatrix}, \mathbf{I} = \begin{bmatrix} I_a \\ I_b \\ I_c \end{bmatrix}, \mathbf{R} = \begin{bmatrix} R_a & 0 & 0 \\ 0 & R_b & 0 \\ 0 & 0 & R_c \end{bmatrix} \qquad \text{Eq. 2.8}$$

Re-writing Eq. 2.2 in matrix notation:

$$\mathbf{E} = \frac{d}{dt}\boldsymbol{\lambda} \qquad \text{Eq. 2.9}$$

Given that λ is defined in Eq. 2.10:

$$\boldsymbol{\lambda} = \begin{bmatrix} \lambda_a \\ \lambda_b \\ \lambda_c \end{bmatrix} \qquad \text{Eq. 2.10}$$

Re-writing Eq. 2.3 in matrix form:

$$\begin{bmatrix} \lambda_a \\ \lambda_b \\ \lambda_c \end{bmatrix} = - \begin{bmatrix} L_a & L_{ab} & L_{ac} \\ L_{ba} & L_b & L_{bc} \\ L_{ca} & L_{cb} & L_c \end{bmatrix} \begin{bmatrix} I_a \\ I_b \\ I_c \end{bmatrix} + \begin{bmatrix} L_{af}(\theta_e) \\ L_{bf}(\theta_e) \\ L_{bf}(\theta_e) \end{bmatrix} I_f \qquad \text{Eq. 2.11}$$

Based on Eq. 2.11, magnetic flux of the individual phases $\boldsymbol{\lambda}$ is finally defined in Eq. 2.12

$$\boldsymbol{\lambda} = -\mathbf{LI} + \mathbf{L}_f^T(\theta_e) I_f \qquad \text{Eq. 2.12}$$

The magnetic flux of the excitation field defined in Eq. 2.4, can be also defined in matrix notation according to Eq. 2.13:

$$\lambda_f = -\boldsymbol{L}_f^T(\theta_e)\mathbf{I} + L_f I_f \qquad \text{Eq. 2.13}$$

For this analysis, it is assumed that the machine is electrically balanced, meaning that the balance equation for the 3-phase currents is:

$$I_a + I_b + I_c = 0 \qquad \text{Eq. 2.14}$$

To simplify the mathematical model of the synchronous machine, a balanced 3-phase circuit is assumed according to the following:

$$R_a = R_b = R_c = R_s \qquad \text{Eq. 2.15}$$

$$L_a = L_b = L_c = L_s$$

$$L_{ab} = L_{ba} = L_{ac} = L_{ca} = L_{bc} = L_{cb} = L_{ss} \qquad \text{Eq. 2.16}$$

3-Phase Diode Bridge Rectifier

The rectifier bridge is modeled based on the switching functions $[g_a, g_b, g_c]$ which are associated with each branch of the bridge. These functions mainly represent the conduction state of the respective diode in the respective bridge-branch, regardless of whether it is active or not [51].

The mathematical representation of the 3-phase diode rectifier is described in terms of V_{dc} in Eq. 2.17 and in terms of I_{dc} in Eq. 2.18

$$V_a = f_a V_{dc} \qquad V_b = f_b V_{dc} \qquad V_c = f_c V_{dc} \qquad \text{Eq. 2.17}$$

$$I_{dc} = g_a I_a + g_b I_b + g_c I_c \qquad \text{Eq. 2.18}$$

and f is defined as:

$$\mathbf{f} = \mathbf{pg} \qquad \text{Eq. 2.19}$$

In more detail:

$$\mathbf{f} = \begin{bmatrix} f_a \\ f_b \\ f_c \end{bmatrix}, \qquad \mathbf{p} = \begin{bmatrix} \frac{2}{3} & -\frac{1}{3} & -\frac{1}{3} \\ -\frac{1}{3} & \frac{2}{3} & -\frac{1}{3} \\ -\frac{1}{3} & -\frac{1}{3} & \frac{2}{3} \end{bmatrix}, \qquad \mathbf{g} \begin{bmatrix} g_a \\ g_b \\ g_c \end{bmatrix} \qquad \text{Eq. 2.20}$$

$$f_a = \frac{2g_a - g_b - g_c}{3} \qquad f_b = \frac{2g_b - g_c - g_a}{3} \qquad \text{Eq. 2.21}$$

$$f_c = \frac{2g_c - g_a - g_b}{3}$$

From Eq. 2.17, the individual phase voltages in vector notation are defined in terms of the output DC voltage V_{dc} of the bridge rectifier according to Eq. 2.22:

$$\mathbf{V} = \mathbf{f}V_{dc} \qquad \text{Eq. 2.22}$$

Similarly, based on Eq. 2.18, the phase currents are formulated in Eq. 2.23 :

$$I_{dc} = \mathbf{g}^T \mathbf{I} \qquad \text{Eq. 2.23}$$

$$\mathbf{I} = (\mathbf{g})^{-1} I_{dc} \qquad \text{Eq. 2.24}$$

Voltage Regulator

It has the primary function of regulating the output voltage of the alternator and maintaining it at a pre-determined level against the full range of engine speed and regardless of the electric load I_L. This voltage is defined by the on-board electrical system of the vehicle. For this particular alternator, the regulated excitation field voltage V_f is pulse-width modulated based on a Proportional Integral (PI) controller that determines its duty cycle. The duty cycle DT of the field voltage is defined in Eq. 2.25:

$$DT = K_p \left(V_{ref} - V_{dc}\right) + K_I \int_{t_0}^{t} \left(V_{ref} - V_{dc}\right) .dt \qquad \text{Eq. 2.25}$$

Based on Eq. 2.25, the excitation field voltage V_f is defined according to Eq. 2.26:

$$V_f = DT\,(V_{dc}) \qquad \text{Eq. 2.26}$$

The Detailed Mathematical Model of the Synchronous Alternator

The **rotor-stator dynamics** by substituting Eq. 2.7, Eq. 2.9 and Eq. 2.12 according to Eq. 2.27:

$$\dot{\mathbf{I}} = -\mathbf{L}^{-1} (\mathbf{RI} + \mathbf{V}) + \mathbf{L}^{-1} \left(\frac{d}{d\theta_e} L_f(\theta_e)\, \omega_e I_f + \mathbf{L}_f(\theta_e)\, \dot{\mathbf{I}}_f \right) \qquad \text{Eq. 2.27}$$

The **excitation field dynamics**, from Eq. 2.1, Eq. 2.2 and Eq. 2.13:

$$\dot{I}_f = -\frac{R_f}{L_f} I_f - \frac{1}{L_f} V_f - \frac{1}{L_f} \left(\frac{d}{d\theta_e} L_f(\theta_e)\, \omega_e \mathbf{I} + \mathbf{L}_f^T(\theta_e)\, \dot{\mathbf{I}} \right) \qquad \text{Eq. 2.28}$$

Knowing the parameter values of the above generalized model of the alternator, it is possible to obtain a reliable simulation model with a good accuracy. However, it is difficult to utilize this mathematical model in this particular form to detect faults in the EPGS system caused by damaged components. The reason for this is that the effects of these faults are usually observed on the dynamic behavior of the machine the transient part, as well as, the steady-state of the currents, not on the physical parameter values. Therefore, there is a need for a simplified, equivalent, and parametrized model which describes the dynamic behavior of the alternator with good reliability and at the same time allows for observing and isolating the effect of faults directly on the parameter values.

For this purpose, the following subsection presents a simplification of the alternator model so that the effect of faults can be directly isolated and observed at the parameter values.

2.1.2 Simplified Mathematical Model of the Alternator

In order to establish a foundation for the development and implementation of a model-based approach for fault detection in alternators, a simplified and reliable dynamic model is required. The detailed mathematical model is presented in the previous subsection 2.1.1, having knowledge of the physical parameters provides a deep insight into the dynamic behavior of the electric machine in addition to simulating its behavior with a very good accuracy. On the other hand, this model contains coupled and non-linear ordinary differential equations and it is not possible to observe and isolate the effect of faults within the EPGS system using the values of the physical parameters.

In addition to that, the values of the physical parameters themselves depend strongly on the current operating point of the machine, and therefore a simplified model must be implemented for fault detection purposes which allows isolating the effects of these faults.

For the above mentioned reasons, a linearized and simplified equivalent mathematical model of the alternator is adequate. This equivalent model captures the main dynamic behavior. Model simplification is accomplished on the basis of analyzing the functional similarity between this 3-phase synchronous machine with its bridge rectifier and a DC generator. This simplification model maps the main inputs to the main outputs using parametrized ordinary differential equations and does not necessarily completely describe the physics behind the system.

For model simplification, the **main inputs** are defined as:

- The excitation field Voltage V_f in volt.
- Electrical rotation speed ω_e in rad/s.

and the **main outputs** are:

- The current of the excitation field I_f in A.
- The rectified DC generator current at the output of the bridge I_{dc} in A.

The detailed mathematical model represented by Eq. 2.27 and Eq. 2.28 can be simplified and linearized considering the following aspects:

- Linear approximation around a small change of θ_e as it defines the operating point of the alternator.
- Substituting the switching function of the bridge rectifier from Eq. 2.17.
- Substituting the rectified voltage V_{dc} and rectified current I_{dc} according to Eq. 2.17 and Eq. 2.18.

Details of alternator model simplification are provided in Appendix A.

The simplified and equivalent dynamic model is defined by Eq. 2.29

$$\dot{I}_f = -\alpha I_f + \beta V_f \qquad \text{Eq. 2.29}$$
$$\dot{I}_{dc} = -\gamma I_{dc} + \eta\omega_e + \kappa I_f - \lambda V_{dc}$$

This form of the simplified mathematical model is implemented in [12, 13] and [14] and it is a linear approximation of the 3-phase synchronous machine combined with the 3-phase bridge rectifier as a DC generator machine due to their functional similarity.

The parameters of the simplified model are defined as:

- Parameter α depends on the excitation field resistance and inductance as in Eq. 2.30

$$\alpha = \frac{R_f}{L_f} \qquad \text{Eq. 2.30}$$

- Parameter β depends on the excitation field inductance as in Eq. 2.31:

$$\beta = \frac{\left(1 - R^{-1}D^{-1}\mathbf{L}_f\left(\theta_e\right)\left(\mathbf{g}^T\right)^{-2}\right)}{L_f} \qquad \text{Eq. 2.31}$$

- Parameter κ depends on the stator-rotor mutual inductance as in Eq. 2.32:

$$\kappa = \frac{\left(\mathbf{g}^T\right)\mathbf{L}_f\left(\theta_e\right)R}{1 - \mathbf{L}L_f} \qquad \text{Eq. 2.32}$$

- Parameter γ depends on the rotor resistance and inductance as in Eq. 2.33:

$$\gamma = \frac{\mathbf{L}L_f - \mathbf{L}_f^2\left(\theta_e\right) - L_f\mathbf{R}}{\mathbf{g}^T\left(1 - \mathbf{L}L_f\right)} \qquad \text{Eq. 2.33}$$

- Parameter η depends on the field and stator resistances as in Eq. 2.34:

$$\eta = \frac{\mathbf{dM}I_{dc_0}}{\mathbf{L}L_f - 1} \qquad \text{Eq. 2.34}$$

- Parameter λ depends on the excitation field inductance as in Eq. 2.35

$$\lambda = \frac{\left(\mathbf{g}^T\right)\left(\mathbf{L}_f\left(\theta_e\right)DT - \mathbf{pg}L_f\right)}{\mathbf{1} - \mathbf{L}L_f} \qquad \text{Eq. 2.35}$$

Considering the equations Eq. 2.30 to Eq. 2.35, parameters of the simplified and equivalent mathematical model represented by Eq. 2.29 are strongly dependent on the operating point of the machine. In contrast, this representation facilitates the detection and isolation of fault effects within the alternator machine, as demonstrated earlier in [14].

In order to create a reliable simulation model that simulates the healthy or, in other words, the reference operating state of the alternator machine, the parameter values of the simulation model must be estimated for the complete operating range of the machine. For this purpose and in order to parametrize the simulation model, measurement data representing the operating range of the machine need to be utilized. These data are obtained from a specially designed experimental setup and based on dedicated measurement series and experiments which are presented in the following Section 2.2.

2.2 Experimental Setup and Measurements

The experimental setup is designed by using a test bench that has the alternator machine as the focus of the experiments. Within the test bench the alternator is operated and controlled similarly to those within the vehicle, therefore, the experimental series are designed correspondingly. The subsequent subsections introduce the test bench and present the measurement data.

2.2.1 Test Bench Setup

Experimental data for parameterizing the mathematical model of the alternator are obtained from a test bench specially designed for this purpose. The experimental setup is designed and constructed on the "High-Speed Driving Machines" Test bench at the FKFS Institute. Figure 2.3 presents a schematic diagram of the constructed experimental setup.

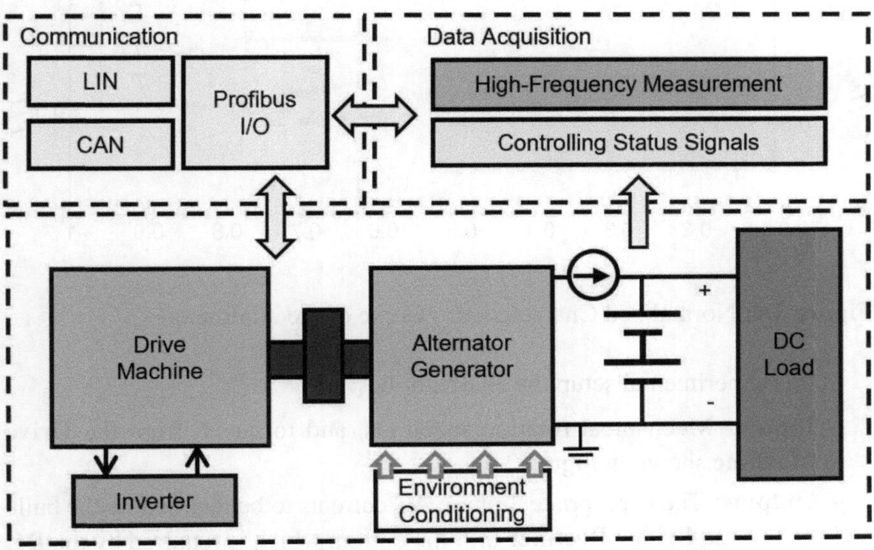

Figure 2.3: Diagram of the Experimental Setup

Outlined below, with reference to this schematic, is a concise description of the key elements of this experimental setup, together with the primary measurement signals and sensors:

- **Alternator**: is the device under test from which its mathematical model is derived and where the main approach of this dissertation is implemented. The automotive machine is a 3-phase synchronous machine and is typically used in small to medium commercial vehicles.
  Figure 2.4 illustrates the normalized characteristics curve of the alternator. The figure clearly demonstrates that the alternator reaches its maximum η_{eff} of about 70 % at around 15 % of its maximum speed and 50 % of its maximum output current I_{dc}. This characteristic allows the machine to deliver comparatively high currents at relatively low speeds. In addition to that, the machine itself requires relatively low torque and the electric power-to-weight ratio is relatively high with long durability and high performance. The machine has a nominal power of 4500 W.

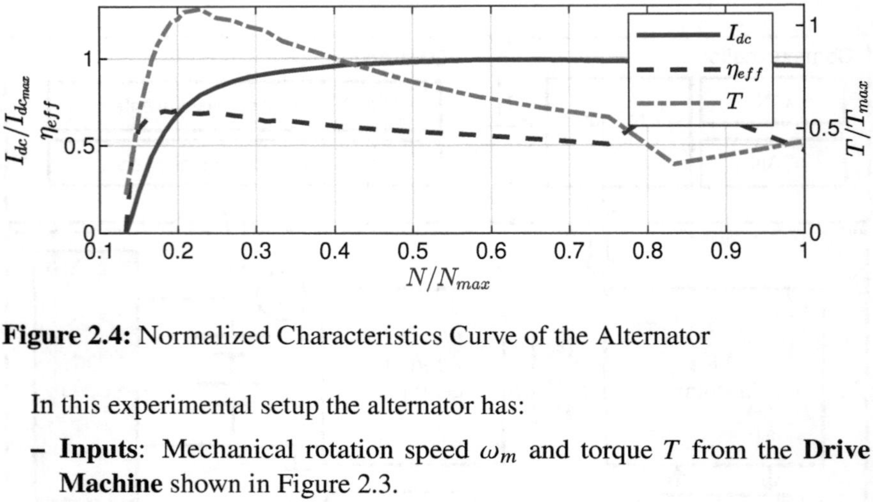

Figure 2.4: Normalized Characteristics Curve of the Alternator

  In this experimental setup the alternator has:

  - **Inputs**: Mechanical rotation speed ω_m and torque T from the **Drive Machine** shown in Figure 2.3.
  - **Outputs**: The appropriate 3-phase AC currents to be rectified by the built-in 3-phase **Bridge Rectifier** into I_{dc} current which is required by the **DC Load**.
  - **Control**: Based on communication via a LIN bus. The controller determines the duty cycle of V_f based on the deviation between V_{dc} and the

predefined reference voltage of the vehicle electric system, which is 28.5 V for this particular case.

A detailed list of the measurement signals and the utilized sensors is provided in Table 2.1.

Table 2.1: Measurement Signals of Experimental Setup

Signals	Sensor	Device
I_{dc}	Open-loop Hall effect-based	Alternator
I_a, I_b, I_c	Electrostatic shielded Rogowski compact coil	
V_a, V_b, V_c	Oscilloscope differential probes	
V_{dc}	Directly on the data acquisition system using a sensor with common ground	
V_f	Directly on the data acquisition system using a sensor with common ground	
I_f	Open-loop Hall effect-based	
ω_m	Encoder	Drive Machine
T	Calculated by machine controller	
I_L	Open-loop Hall effect-based	DC Load
V_L	Directly on the data acquisition system using a sensor with common ground	
I_{Bat}	Open-loop Hall effect-based	Battery
V_{Bat}	Open-loop Hall effect-based	

- **Drive Machine**: Is a synchronous machine with high power and torque per weight ratio. This machine is suitable for high-speed applications, has minimum drag losses, and speed is controlled via CAN bus by an inverter.
- **DC Load**: Is a programmable DC load with energy recovery to the grid and is controlled via the CAN bus. The DC load receives a maximum electric power of 7 kW, maximum voltage of 80 V and maximum current of 340 A. The internal-controlled variable resistance varies between $[0.005, 6]\ \Omega$.
  - **Inputs**: The required load current I_L at a constant voltage of V_{dc}.
  - **Outputs**: AC Voltage of 230 V with frequency of 50 Hz and that is the standard European grid specification.

- **Battery**: Supplies the required power when the alternator cannot provide it fully, as well as the necessary starting power. The system contains two 12 V batteries connected in series, each with a capacity of 10 Ah and a maximum current of 150 A Cold Cranking Amps (CCA).
  - **Inputs**: Battery current I_{Bat} during charging.
  - **Outputs**: Battery current I_{Bat} during discharging.

  The battery itself is based on an absorbent-Glass Mat (AGM Lead-Acid technology and is carefully selected for this experimental setup because this technology enables the battery to:
  - Repeatedly charged and discharged without any loss of performance.
  - Handle relatively large electrical loads and regenerative braking.
  - Provide high-starting power.
- **Data Acquisition**: Is a high-speed data acquisition system and communication is via Ethernet. The system is equipped with isolated analog input amplifier modules and offers high bandwidth, relatively high accuracy at different input ranges, and integrated filters. The measurement system is also equipped with interchangeable modules that have a high galvanic isolation. This system offers a wide range of sampling rates. The sampling rate for this experimental setup is set to 12.5 kHz.

By utilizing the above described experimental setup, measurement data with a suitable sampling rate are acquired and implemented to estimate the values of parameters within the alternator model. The following section presents these measurement data and the series of experiments.

2.2.2 Experimental Data

In order to estimate values of the parameters within the mathematical model of the alternator that results in a reliable and representative simulation model, the input and output signals of the alternator within the experimental setup described in Subsection 2.2.1 are measured within a relatively large number of series of experiments.

The measurement data are required to cover the main operating range of the alternator and capture its most important dynamics with good accuracy. As a matter of course, the accuracy of the model depends on the quality of the measured signals, which in turn depends on the experimental setup. In addition to accuracy of the measured data, the duration of the measurement series need to be sufficient to capture the main time constants of the model and thereby capturing the complete dynamic behavior including the transient as well as the steady-state.

The experiments have been planned considering the characteristic curve of the alternator as shown in Figure 2.4. They are planned according to two primary configurations:

- A step change in the electric load in increasing and decreasing order at several constant rotational speeds covering a representative operating range of the alternator. These experiments are shown in Figure 2.5. The step change in the electric load is $\Delta I_L = 25\,\text{A}$ and a steady-state time of $t_s = 60\,\text{s}$ are designed so that the inputs adequately excite the system dynamics and allow enough time to capture the main time constants.

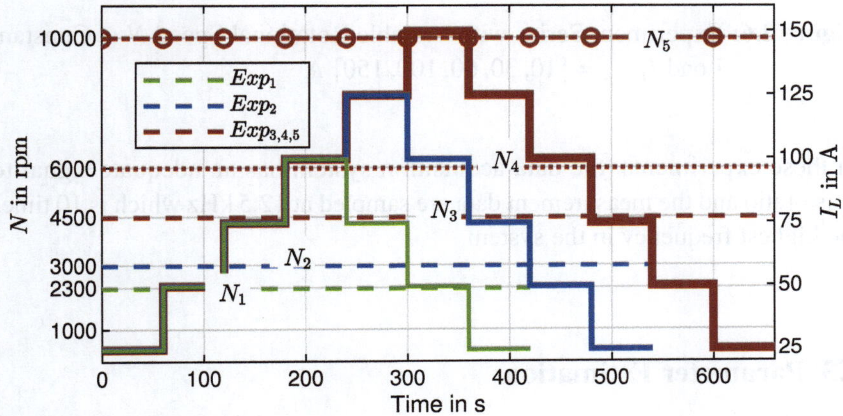

Figure 2.5: Experiment Series with Variable I_L at Constant Rotational Speeds $N_{1\cdots5} = [2300, 3000, 4500, 6000, 10000]$ rpm

- A step change in the rotational speeds in increasing and decreasing order at constant electric loads covering a representative range of electric loads of the alternator. These experiments are shown in Figure 2.6. The step change in the rotational speed is $\Delta\omega = 500$ rpm and a steady-state time of $t_s = 10$ s are designed to capture the effects of changing rotational speed on the electric load.

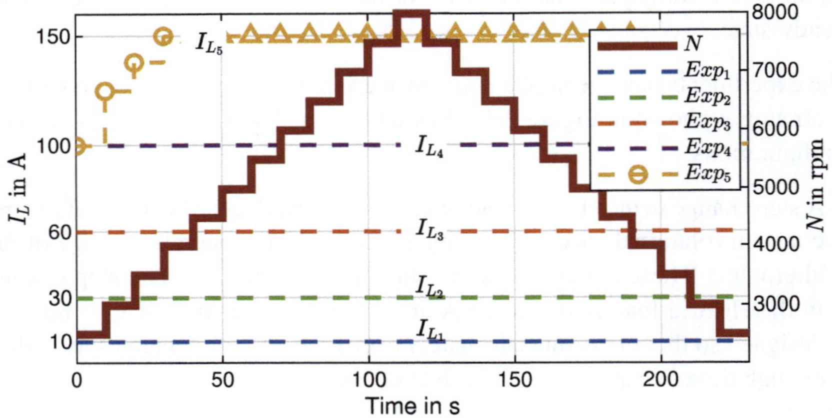

Figure 2.6: Experiment Series with Variable Rotational Speed N at Constant Load $I_{L_{1\cdots5}} = [10, 30, 60, 100, 150]$ A

In these experiments, the data acquisition system has an adequate signal-to noise-ratio and the measurement data are sampled at 12.5 kHz which is 10 times the highest frequency in the system.

2.3 Parameter Estimation

Parameter estimation is the process of computing the values of model parameters from measured data. Its main goal is to find the values that maximize the agreement between the model predictions or outputs and the hardware-obtained measurement data. Maximizing the agreement can be interpreted as minimizing the deviation between the measurement and the model output. Figure 2.7 illustrates a schematic diagram of the parameter estimation process.

Parameter estimation is applied to different types of mathematical models; including statistical models, parametric dynamic models, and empirical data-based models. In this work, parameter estimation is applied to a linearized and equivalent mathematical model of the alternator which describes the dynamics of the machine.

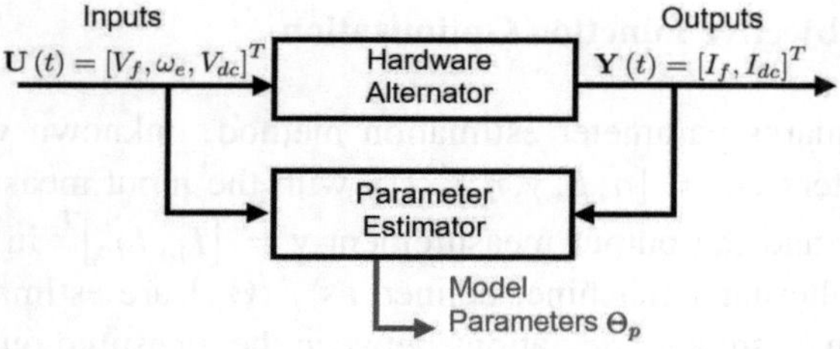

Figure 2.7: Parameter Estimation Based on Measurement Data

The parameters of the alternator model defined by Eq. 2.29 and as can be clearly seen from Eq. 2.30 to Eq. 2.35 are highly dependent on the operating condition of the machine. In this context, the parameter estimation scheme provides parameter values according to each operating point of the machine or, in other words, with respect to electric load and rotational speed. Then with the help of lookup tables, it is possible to assign the proper individual value of each parameter based on the current operating point.

Minimizing the deviation between the measurement data and the output of the model requires a function that describes it. This function, which is implemented within the estimator shown in 2.7 needs to be gradually and iteratively minimized by adjusting the values of the model parameters. In this work, the function is defined as the objective function or the cost function within an optimization problem. The most common approaches for parameter estimation are the following:

- Least squares.
- Maximum likelihood.

- Bayesian statistics.

In this dissertation, the least squares method is implemented to obtain parameter values of the simplified and linearized model of the alternator. The reason for this choice is the ease of implementation and efficiency in addition to providing good statistical properties. This estimator is utilized within a multi-objective optimization problem which will be presented in the following subsection 2.3.1.

2.3.1 Multi-Objective Function Optimization

In the least squares parameter estimation method, unknown values of the model parameters $\Theta_p = [\alpha, \beta, \gamma, \eta, \kappa, \lambda]^T$ with the input measurement $\mathbf{u} = [V_f, \omega_e, V_{dc}]^T$ and the output measurement $\mathbf{y} = [I_f, I_{dc}]^T$ in a regression model of the alternator machine, defined as $f(\Theta_p)$ are estimated by minimizing the sum of squared deviations between the measured outputs and the simulated outputs of the simulation model. Considering the output of the model in Figure 2.7 is defined as in Eq. 2.36 with m parameters:

$$y = u_1\theta_{p_1}+, ..., +u_m\theta_{p_m} \qquad \text{Eq. 2.36}$$

Assuming that each measurements series has n number of samples, then the output of a generalized regression model is expressed according to Eq. 2.37:

$$\begin{bmatrix} y_1 \\ \vdots \\ y_n \end{bmatrix} = \begin{bmatrix} u_{1,1} & \dots & u_{1,m} \\ \vdots & & \vdots \\ u_{n,1} & \dots & u_{n,m} \end{bmatrix} \begin{bmatrix} \boldsymbol{\theta}_{p_1} \\ \vdots \\ \boldsymbol{\theta}_{p_m} \end{bmatrix} = \begin{bmatrix} \mathbf{u}_1 \\ \vdots \\ \mathbf{u}_n \end{bmatrix} \begin{bmatrix} \boldsymbol{\theta}_{p_1} \\ \vdots \\ \boldsymbol{\theta}_{p_m} \end{bmatrix} \qquad \text{Eq. 2.37}$$

The deviation or accordingly the error vector between the measured output and the simulated output of the model is:

$$\mathbf{Er} = \begin{bmatrix} e_1 \\ \vdots \\ e_n \end{bmatrix} = \begin{bmatrix} y_1 - \mathbf{u}_1\Theta_p \\ \vdots \\ y_n - \mathbf{u}_n\Theta_p \end{bmatrix} \qquad \text{Eq. 2.38}$$

The error vector:

$$\mathbf{Er} = \mathbf{Y} - \mathbf{U}\Theta_p \qquad \text{Eq. 2.39}$$

The problem is defined to find as estimate of the unknown parameter set defined by Θ_p, so that the cost function $\mathbf{Er}^T\mathbf{Er}$ is minimized according to the optimization problem in Eq. 2.40:

$$\begin{aligned}\min_{\theta_p} \|\mathbf{Er}\|_2^2 &= \min_{\theta_p} \mathbf{Er}^T\mathbf{Er} \\ &= \min_{\theta_p} \sum_{t=0}^{n} \left(e\left(\theta_p\right) \times e\left(\theta_p\right)\right)\end{aligned} \quad \text{Eq. 2.40}$$

Recalling that the simulation model of the alternator has two outputs I_f and I_{dc}, the optimization problem in Eq. 2.40 is re-formulated to include multi-objective functions according to:

$$\min_{\theta_p} \|\mathbf{E}\|_2^2 = \min_{\theta_p} \sum_{t=0}^{n} \left(e_{I_f}\left(\theta_p\right) \times e_{I_f}\left(\theta_p\right) + e_{I_{dc}}\left(\theta_p\right) \times e_{I_{dc}}\left(\theta_p\right)\right) \quad \text{Eq. 2.41}$$

The following section presents results of parameterizing the simplified alternator model in addition to the simulation results.

2.3.2 Results Simulation Model

This section presents the results of the implementation of the method described in Subsection 2.3.1 for parameter estimation using multi-objective and multi-starting point optimization. The measurement data are obtained from the series of experiments mentioned in Subsection 2.2.2.

Parameter values as mentioned earlier are highly dependent on the operating point of the machine therefore, for each parameter within the alternator model described by the Eq. 2.29, a set of values is obtained based on measurement data of experiments characterizing the respective operating points. Based on this, a representing operating range of the alternator is defined which is for I_L between (25 to 150) A and rotational speed between (2500 to 10 000) rpm, then this range is divided into several sub-operating points. These operating points are defined according to the following segmentation of the measurement series:

- At every constant speed of $N = [3000, 4500, 6000, 10000]$ rpm an increasing and decreasing steps of 25 A of I_{dc} is segmented from the measurement series and used to estimate the corresponding model parameters. This is achieved for the whole range of I_{dc} between (25 to 150) A.

- At every constant current of $I_{dc} = [10, 30, 60, 100, 150]$ A an increasing and decreasing speed steps of 500 rpm between (2500 to 8000) rpm.

The operating point at $N = 3000$ rpm and I_{dc} in the range (25 to 50) A is selected to demonstrate results of the parameter estimation algorithm. This operating range of the alternator is selected based on the analysis provided from the driving route in [52]. In this research, the measurement data for a similar alternator is provided within a medium commercial vehicle.

The estimated parameter set Θ_p is implemented within the simulation model and the results are compared with the measurement data in Figure 2.8. It can be clearly seen that the courses of simulation data match the measurement with very good accuracy. The transient behavior represented by the several overshoots and steady-state before and after load change, matched with good accuracy for I_{dc}. A slight deviation in I_f of 10 % from the maximum overshoot and a delay of less that 0.2 ms is registered. This deviation is related to the structure of the optimization problem as it minimizes multi-objective functions and the final solution provides the minimum as a trade-off between all the objective functions, in other words the residuals of I_{dc} and I_f. In this context, more weight for I_{dc} is assigned because it corresponds to the electric load required by the vehicle.

For this operating range, the trajectories of the parameter values in each iteration during the search for the minimum of the optimization problem are shown in 2.9a. Here, it is possible to observe that the algorithm settles at the global minimum after 60 iterations with estimated parameter values of $\Theta_p = [11.3, 1.4, 664, 1490, 94140, 8870]^T$. The course of the change of the cost function as being minimized can be seen in 2.9b where it settles at the minimum. In order to evaluate the reliability and performance of the simulation model that contains the complete set of parameter values and covers the representative or dominant operating range of the alternator machine, the measurement data of the driving route provided in [52] are applied. In this study, the average energy consumption of the commercial vehicle on German roads is analyzed and optimized. For this purpose, a special driving route around Stuttgart city consists of the four main types of road in Germany. The purpose

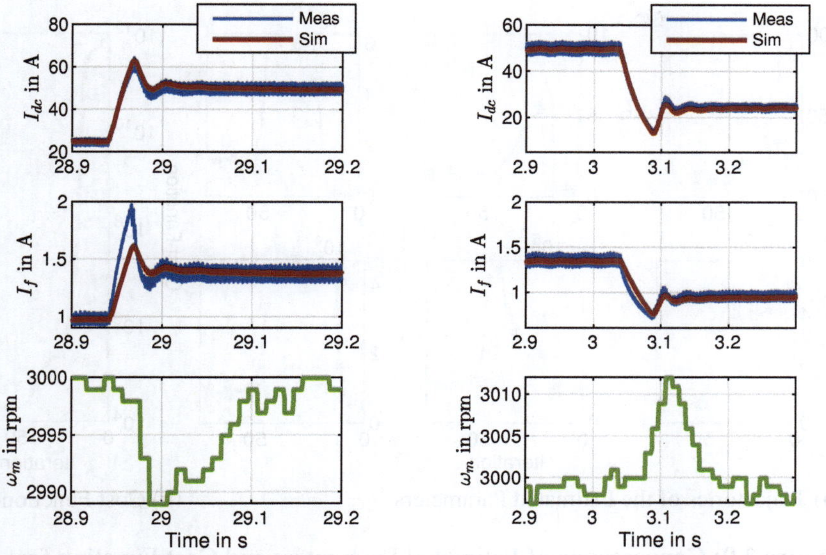

(a) Increasing I_{dc} from (25 to 50) A **(b)** Reducing I_{dc} from (50 to 25) A

Figure 2.8: Simulation Results for $I_L \in [25, 50]\,A$ at Constant N =3000 rpm

of this study is to analyze the influence of the driver on the average energy consumption during driving.

This test vehicle is equipped with an equivalent alternator; therefore, its operation during this route can be used to evaluate the performance of the developed simulation model. The former is accomplished by using measurement data of the rotational speed and electric load. Figure 2.10 presents a 10 s segmentation of the results of the simulation model. In this simulation, electric load requirements are applied in addition to engine speed during driving to the simulation model. It can be clearly seen that the output of the simulation model represented by the electric load follows the required load provided by the measurement with very good accuracy. The on-board voltage is also regulated around a reference value of 28.5 V and the maximum deviation is around ±0.3 % indicating a successful performance.

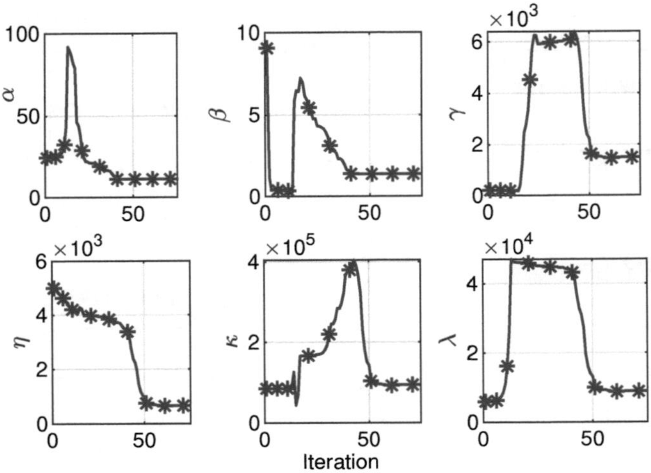

(a) Trajectories of the Estimated Parameters

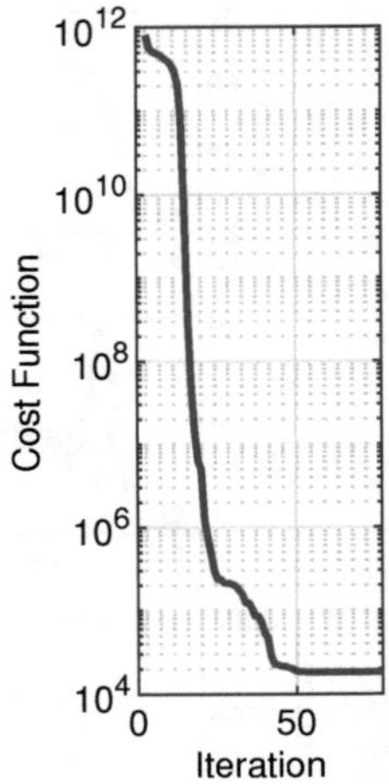

(b) Cost Function

Figure 2.9: Convergence of Estimated Parameters and Cost Function for $I_L \in [25, 50]A$ at Constant N =3000 rpm

In addition to the performance, it is adequate to examine the values of model parameters as they change according to the corresponding operating range. The course of change of parameter values during this simulation segmentation is presented in Figure 2.11. The figure demonstrates the smooth transition of all parameter values derived from their respective lookup tables, in alignment with different operational points designated by the measurement data of the planned route.

Based on the above, the reliability of the simulation model has been validated. Therefore, it can be used for further development and testing of the model-based fault detection and compensation method. The next chapter presents the fault detection method that utilizes this simulation model and is based on the deviation of the parameters.

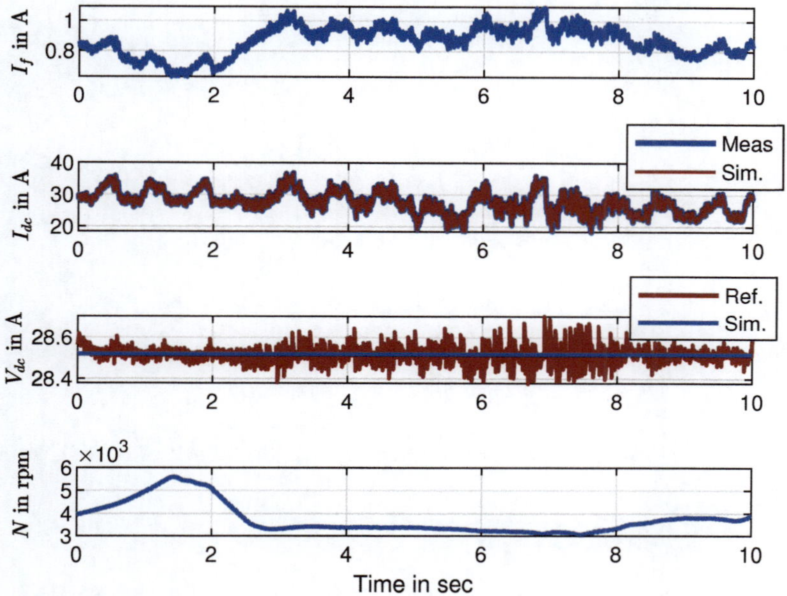

Figure 2.10: Segmentation of Simulation Results for a Driving Route

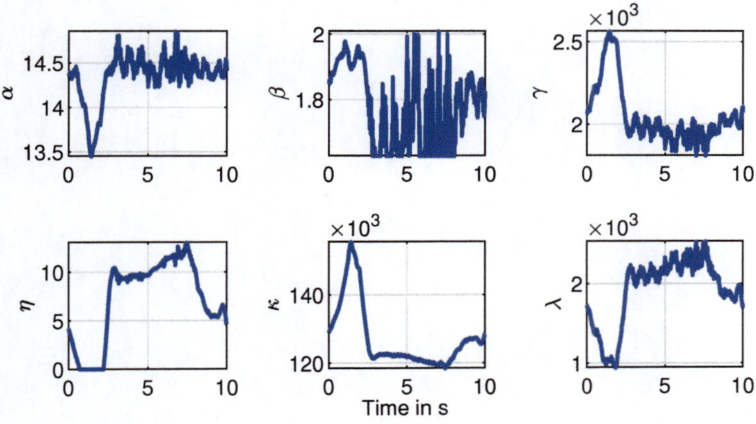

Figure 2.11: Parameters Transitions During a Segmentation of the Driving Route

3 Model-Based Fault Detection for Automotive Electric Machines

Model-based Fault Detection (FD) is an approach that utilizes mathematical models of the system or device being monitored to detect, identify, and diagnose potential faults. These models can be based on physical laws, engineering principles, or data-driven approaches. As faults may change the behavior and performance of the system. These changes can be utilized to indicate inherent faults that are not measurable and cannot be located prior to the total failure of the system or component.

In general, Fault Detection and Identification (FDI) is classified in the literature mainly into model-based and non-model-based approaches, each classification has its own line of methods according to Figure 3.1 as obtained from [23]:

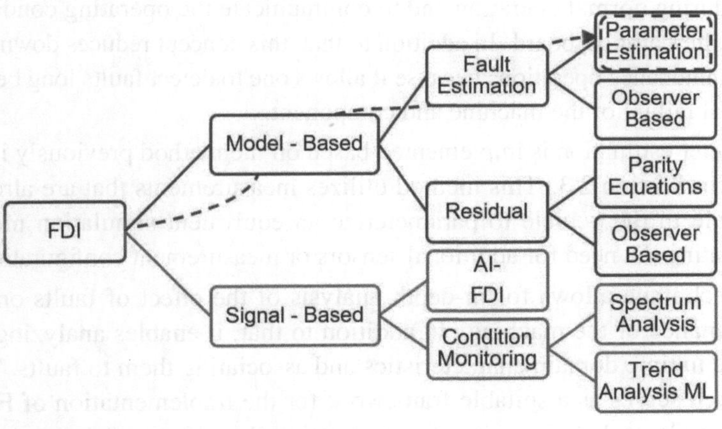

Figure 3.1: Classification of FDI Types

This dissertation introduces a model-based approach for FD for automotive electric machines. The implementation is applied to an alternator as an example, but is not limited to. This technique employs parameter estimation as a

A. Salah, *Model-Based Fault Diagnosis and Fault-Tolerant Control*,
Wissenschaftliche Reihe Fahrzeugtechnik Universität Stuttgart,
https://doi.org/10.1007/978-3-658-50108-2_3

basis for detecting faults. The main goal of this implementation is to derive a parametrized model which represents the healthy machine or the reference operating state in addition to a model which represents the actual machine-with-faults. Comparing both sets of parameters allows for the detection of faults if any deviation or change in their values exceeds a pre-defined threshold relative to the reference parameter set.

This implementation of model-based FD in the automotive domain has several advantages compared to the other approaches, which can be summarized by the following:

- In addition to fault detection, this approach provides a model of the reference machine which runs within an assigned cloud parallel to the actual hardware. This implementation over the cloud or on a high-power computational server is considered a realization of Digital Twin (DTw) of the machine in operation.
- Exploiting the assigned DTw, this enables the vehicle to perform self-diagnosis during normal operation and to communicate the operating condition of the alternator on-board. In addition to that, this concept reduces downtime and maintenance operations because it allows one to detect faults long before the total failure of the machine and component.
- Parameter estimation is implemented based on the method previously introduced in Section 2.3. This method utilizes measurements that are already available in the vehicle to parameterize an equivalent simulation model, eliminating the need for additional sensors or measurement configurations.
- This technique allows for in-depth analysis of the effect of faults on the performance of the machine. In addition to that, it enables analyzing the change in time domain characteristics and associating them to faults. This approach serves as a suitable framework for the implementation of Fault-Tolerant Control (FTC), aimed at mitigating the effects of faults on the performance of the machine and recovering its reference behavior.

The following Section 3.1 presents common faults in alternators and investigates the effects of these faults on the performance in addition to the change in its time domain characteristics. Then Section 3.2 introduces the proposed method for model-based FD and its simulation results.

3.1 Faults in Automotive Electric Machines

According to International Federation of Automatic Control (IFAC) technical committee as outlined in [53], a fault is defined as an un-permitted deviation of at least one characteristic, property, or parameter of the system from the acceptable, usual or standard conditions. When faults occur in a system or actuator, they can still operate with less effective performance. This section introduces the common faults that occur in automotive electric machines and highlights the effect of these faults on the performance and time domain characteristics.

3.1.1 Common Faults

There are many faults that affect the performance and operation of alternators as well as traction motors in vehicles. A considerable number of these faults occur mainly but are not limited to due to manufacturing defects such as short circuit (inter-turn) and open circuits in the stator or rotor windings. However, most of these faults occur during the operation of the electric machine. These faults can be electrical as well as mechanical, or even in the control signal. Figure 3.2 summarizes the most common faults in automotive alternators.

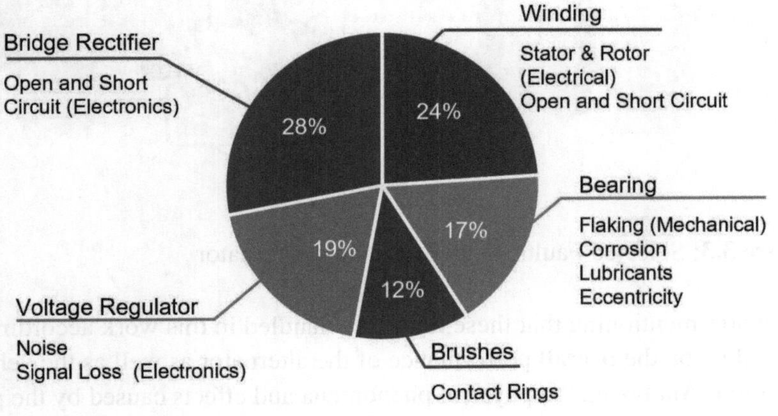

Figure 3.2: Common Faults in Automotive Alternators

The percentage of occurrence of these faults is only an indicator of their dominance over other faults, however, the statistics are different according to several research studies as in [54] and [13] and many others according to [55]. In this dissertation, the following common faults, which are shown in Figure 3.3, are thoroughly considered and analyzed using test bench experiments.

- Mechanical faults in the bearing at both sides of the alternator; the driving side as well as the shaft side. Bearing faults include:
  - Flaking in the rolling element.
  - Corrosion.
- Control signal faults: due to broken strings within the conducting wires. This fault causes an intermittent loss in the control signal of the voltage regulator or produces considerable superimposed noise.
- Damaged diodes in the bridge rectifier.

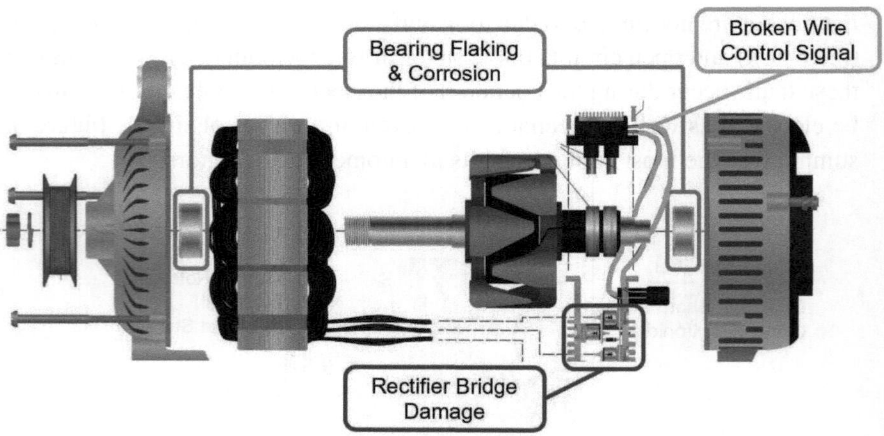

Figure 3.3: Selected Faults on an Exemplary Alternator

It is worth mentioning that these faults are handled in this work according to their effect on the overall performance of the alternator as well as the vehicle operation. Analyzing the physical phenomena and effects caused by the presence of these faults is beyond the scope of this dissertation. For this reason, the next subsection presents the resulting effects of these faults on the efficiency of the alternator, which represents the steady-state. In addition, the characteristics of the machine in the time domain and the dynamic behavior are thoroughly analyzed, as well as the supply of electric load during vehicle operation.

3.1.2 Effects of Faults on the Performance of the Alternator

In order to analyze the effects of the above mentioned common faults on performance of the alternator, it is beneficial to examine the performance of the machine within the selected and representative operating range. This performance, which will be considered as the reference performance of the healthy alternator, is shown in Figure 3.4. On the left side in Figure 3.4a the efficiency of the alternator within the selected operating range versus the output power and rotational speed is illustrated. However, the performance of the machine varies depending on the operating point; it can be seen that the maximum electrical power provided within the selected range is 4500 W with a η_{eff} of 69 %.

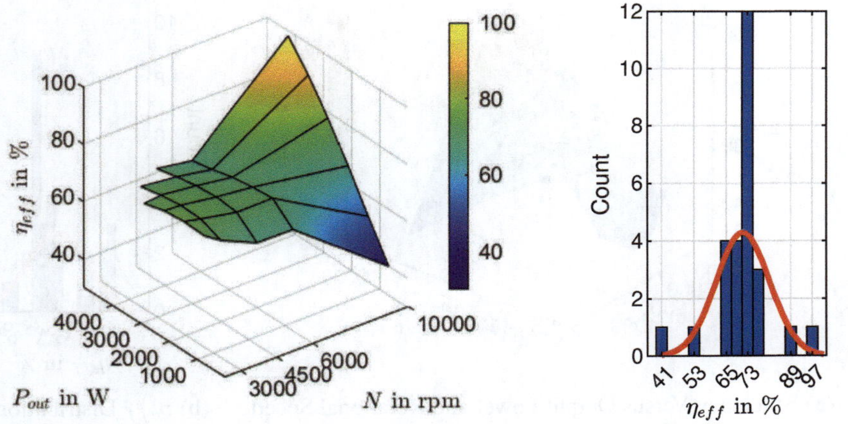

(a) Efficiency Versus Output Power and Rotational Speed **(b)** η_{eff} Distribution

Figure 3.4: Reference Performance of the Alternator within a Selected Operating Range

By considering Figure 3.4b, this efficiency value has a relatively high incidence number. The average efficiency of around 73 % which has the highest incidence number is provided in the range of electric power of P_{out} = (2140 to 2850) W corresponding to I_L = (75 to 125) A and rotational speed of N = (2300 to 5500) rpm. The lowest efficiency in the range from (40 to 60) % is provided at electric power P_{out} = (714 to 1400) W corresponding to I_L = (25 to 50) A and rotational speed N = (6000 to 10 000) rpm which are the highest in the selected operating range.

By evaluating the same alternator that has a corroded ball bearing **Corrosion Fault (CO)** on the driving side, this fault is one of the most common problems of a similar alternator. The performance of the machine with corrosion fault is illustrated in Figure 3.5. It can be clearly seen that the efficiency of the machine is reduced and specially at medium to high electric load and at medium rotational speed of N =(3000 to 5000) rpm. This is represented by the dark blue area in Figure 3.5a at these speeds.

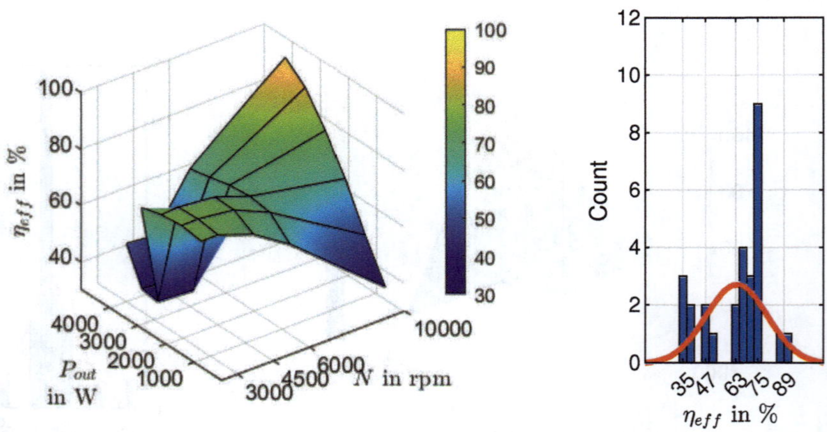

(a) Efficiency Versus Output Power and Rotational Speed **(b)** η_{eff} Distribution

Figure 3.5: Performance of the Alternator with Corroded Ball Bearing

At higher speeds the efficiency is also reduced in general and the peak of the normal distribution curve in Figure 3.5b has been shifted to left with a wider base, resulting in a mean value of 60 % compared to 73 % of the healthy machine. The maximum electric power provided by this machine is P_{out} = 4300 W registering a 5 % reduction compared to the healthy one. In addition to that, maximum power is provided with a reduced efficiency of η_{eff} = (46 to 67) % compared to 69 % by the healthy machine, which requires higher rotational speeds of N =(4500 to 6000) rpm.

In addition, the maximum efficiency of 73 % is recorded with a reduced incidence value of 9 compared to 12 by the healthy machine. In general, corroded ball bearing has resulted in a reduction in the efficiency of the alternator of (4 to 35) %. Based on this analysis, it can be concluded that ball bearing corrosion, which is usually not detected earlier than the total breakdown of the machine, has a severe impact on the performance of the alternator during operation and long before the total failure.

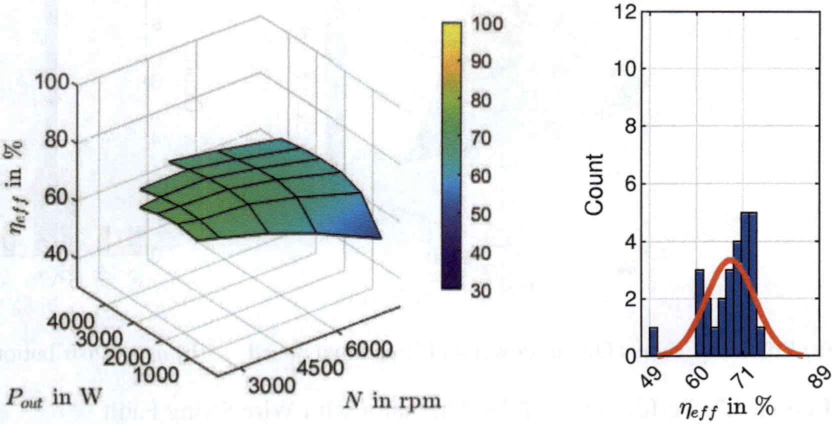

(a) Efficiency Versus Output Power and Rotational Speed **(b)** η_{eff} Distribution

Figure 3.6: Performance of the Alternator with a Ball Bearing Flaking Fault

With regard to **Flaking Fault (FA)** on the rolling element of the ball bearing, the performance of this alternator that contains the fault is shown in Figure 3.6a. The first observation that can be made is the limitation on the operating range of the machine, as it is not possible to operate with rotational speeds higher than N = 8000 rpm. This is due to safety and vibration related issues on the test bench itself. In addition to that, the maximum observed electric power is 4300 W at an efficiency of 66 % at higher speeds of N = (4500 to 6000) rpm.

Examining Figure 3.6b, the average efficiency of the alternator with a flaking fault is shifted to the left by an amount of 2 % and the number of incidences with the highest efficiency is reduced to 5 compared to 12 of the healthy machine. In general, the total reduction in the efficiency of this machine is between (5 to 35) %. Once more, the flaking fault has a noticeable effect on the performance of the alternator; however, it is not detected until the bearing is completely damaged.

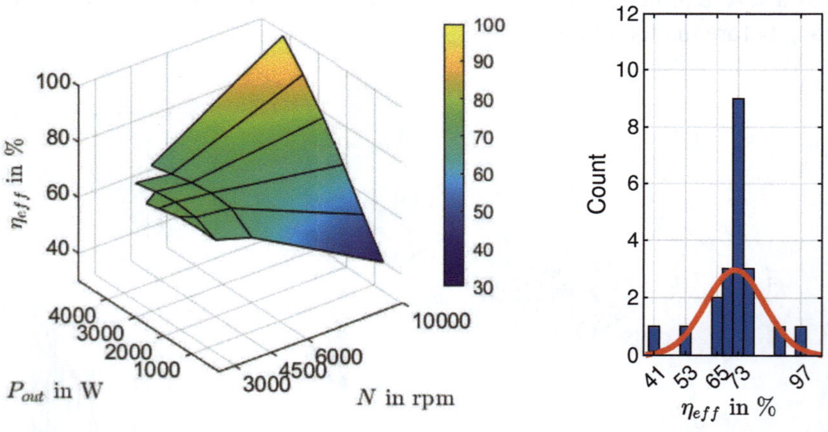

(a) Efficiency Versus Output Power and Rotational Speed **(b)** η_{eff} Distribution

Figure 3.7: Performance of the Alternator with Wire String Fault

The fault of the **Wire String Fault (WS)** that is responsible for the transfer of the voltage regulator control signal is not expected to have a significant influence on the performance of the alternator in general, and this can be seen in Figure 3.7. This may be referred to the nature of this fault as it affects the control signal of the voltage regulator and produces additional noise. However, the influence of this noise can clearly be observed in the time domain characteristics of the alternator. Despite this issue, a slight reduction of 2 % in the efficiency of the alternator can be observed. The highest efficiency has a slightly reduced incidence of 9 compared to 12 and the average value remains at 70 %.

For the **Diode Fault (DF)**, as expected, the effect on the performance is dramatic. The allowable output power that can be generated is reduced to 50 % of the maximum value, and the amount of voltage ripple due to the damaged branch of the bridge rectifier is significant. Figure 3.8 presents the performance of the alternator with damaged diode. The figure illustrates not only the reduced operating range but also a reduction in efficiency by up to 20 % as the electric load increases.

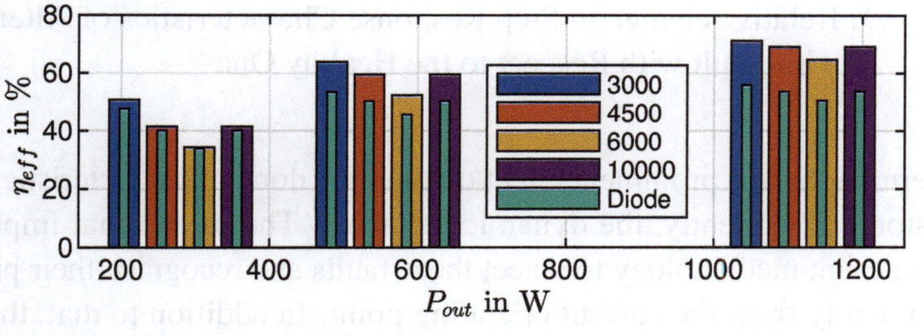

Figure 3.8: Efficiency of Alternator with Diode Faults at Different Loads and Speeds $N_{1\cdots4}$ = [3000, 4500, 6000, 10000] rpm

In addition to the efficiency of the alternator, it is necessary to examine the change in time domain characteristics of the alternator due to faults. Figure 3.9 presents the relative change in the time domain characteristics of the alternator with different types of fault compared to the healthy one. It is observed that there is a remarkable increase in overshoot and settling times with all types of faults and specifically for the alternator with corrosion CO or wire string WS faults. Reduced rise times come at the cost of increased overshoots and extended settling as well as transient times.

It is important to note that these observations regarding the changes in time domain characteristics are dependent on the operating point of the alternator. However, the objective of this analysis is to emphasize the variations in the step response rather than to measure them quantitatively or associate them with various types of fault. The statement here is that these common faults within

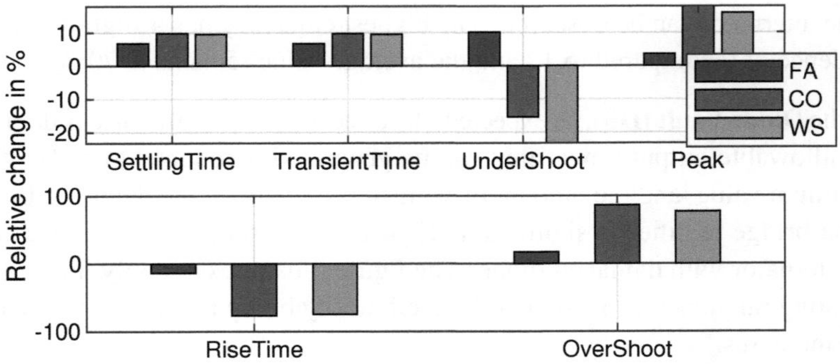

Figure 3.9: Relative change in Step Response Characteristics of Alternator-with-Fault with Respect to the Healthy One

the alternator have a prominent effect on the time domain characteristics of the alternator, consequently, the dynamic behavior. The matter that implies the need to design methodology to detect these faults and recognize their presence independently from the current operating point. In addition to that, there is a need to compensate the degradation in the performance of the alternator using fault tolerant techniques in order to ensure an adequate driving experience.

The following section presents the model-based fault detection approach which takes great advantage of the mathematical model and the relevant information and characteristics.

3.2 Model-Based Fault Detection

The systematic use of the alternator model within the fault detection and compensation method is advantageous. As this model provides significant information about the actual performance of the machine and defines the reference or base-line performance. In addition to that, it is possible to take advantage of this model to generate the digital twin of the machine that runs parallel to the hardware in an assigned cloud or high-performance computational server. The proposed model-based fault detection approach which is introduced in [56] provides a comparison between the reference model or the digital twin of the machine and the actual hardware in real operation. In contrast to the other model-based techniques, model outputs and signals are not used directly in the comparison process, but rather these signals are used in order to estimate the actual parameter values of the model, and then the comparison takes place. In other words, the parameters of the actual system that might contain a fault are compared with the parameters of the reference system. The following section outlines the general framework of the model-based fault detection method and then demonstrates how faults impact the parameter values of the alternator model.

3.2.1 General Framework of Model-Based Fault Detection

It is shown earlier in Section 3.1.2 how the presence of faults has changed the performance of the alternator in steady-state as well as its time domain characteristics. Based on that, it is expected that the parameter values of the derived model of the alternator, which is represented by Eq. 2.29 to change accordingly. Variations in model parameter values due to faults can be leveraged to develop a novel model-based fault detection method by observing how the parameter values differ from those of a healthy or reference model. Figure 3.10 illustrates the general framework of this model-based fault detection method for automotive electric machines.

The following provides a comprehensive description of this framework which is introduced by [57]:

- A reference model which represents the healthy operating state of the machine is mandatory. This reference model is stored either on a designated cloud or on a high-performance computational server and runs in parallel to the actual hardware.
- The alternator of the vehicle is subjected to routine tests designed according to specific driving profiles.
- Measurements of the various variables and the actual response of the machine are transferred to the cloud where the parameter estimator is stored. The parameter estimator is presented earlier in Section 2.3. Moreover, these

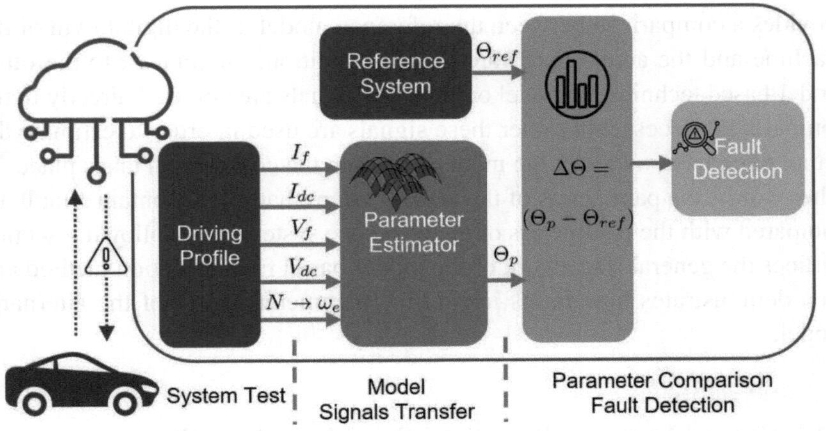

Figure 3.10: Framework of Model-Based Fault Detection for Automotive Electric Machines

measurements that are transferred to parameterize the model are already provided within the vehicle needless to use additional sensors.

- The resulting set of parameters are compared to those of the reference model. Then, whenever a deviation in parameter values exceeds a predefined threshold, this indicates the presence of a fault within the machine.

The subsequent subsection explores the impact of faults on the values of the model parameters, causing them to deviate from the reference model.

3.2.2 Deviation in Model Parameters Due to Faults

In order to analyze the effect of faults on parameter values of the alternator model, the alternator-with-fault that is intentionally introduced is operated in the same series of experiments as those presented in Section 2.2.2. Subsequently, the resulting parameter values are estimated using the estimation problem presented in Section 2.3.1. At this stage, it is of great importance to start from the same initial conditions that have been registered for the reference or healthy machine. In addition to that, the reproducibility of the solution should be guaranteed and the simulation results should be within acceptable accuracy.

The following analysis presents the effects of faults on the parameter values of the alternator model. For simplicity, specific operating points are selected because of their dominance in the driving profile presented by [52]. These points are represented by the rotational speed $N = 3000$ rpm at various electric loads $I_L = [25, 50, 75, 100, 125]\,A$. The results of additional operating points are introduced in the Appendix B.

Initially, the simplified and equivalent alternator model presented in Section 2.1.2 is restated in the following equation Eq. 3.1:

$$\begin{aligned} \dot{I}_{dc} &= -\gamma I_{dc} + \eta\omega_e + \kappa I_f - \lambda V_{dc} \\ \dot{I}_f &= -\alpha I_f + \beta V_f \end{aligned} \qquad \text{Eq. 3.1}$$

The change in α the first parameter within the model due fault is shown in Figure 3.11. One can observe a significant change in α values due to all types of faults, particularly at large electric loads I_L = (100 to 125) A and lower to medium I_L = (25 to 50) A. At these electric loads, the large increase in α up to 130 % due to Wire String Fault (WS) fault is remarkable. This is how this specific fault reveals its effect on I_f. For electric loads in the middle to high range I_L = (50 to 100) A the change in α is slight with a minimum value of 5 %, however, the values maintain the same tendency of change with respect to I_L.

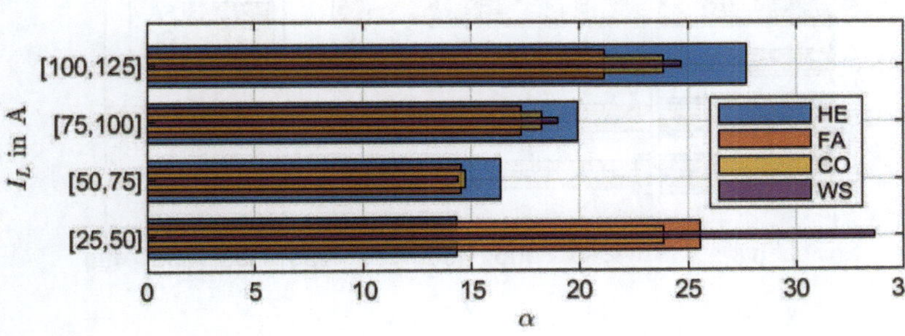

Figure 3.11: Deviation in α Caused by Different Faults at $N = 3000$ rpm

With respect to β, the changes in values due to faults are depicted in Figure 3.12. The change in β reveals the same trend as the change in α with a maximum

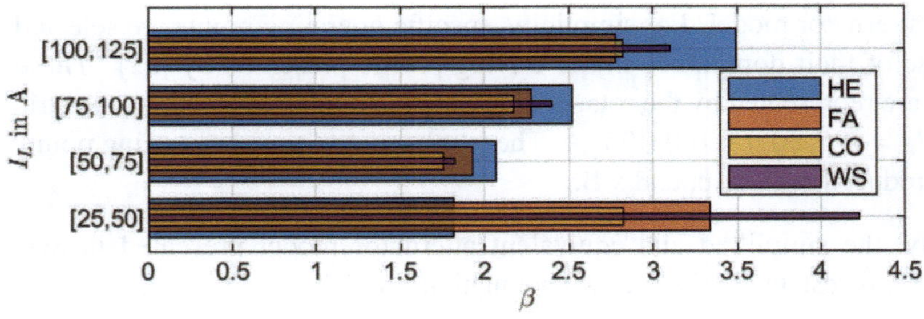

Figure 3.12: Deviation in β Caused by Different Faults at $N = 3000$ rpm

increase of 130 % at the load current I_L =(25 to 50) A and the minimum of 5 % in the range of (50 to 100) A.

For the parameter η which is the factor multiplied by the rotational speed ω_e, the pattern is completely different. With respect to the healthy alternator, η varies significantly as the electric load I_L varies, therefore, it is expected to observe the same pattern for η in the models of alternators with different faults. Also, the range of change in η is notably high and wide.

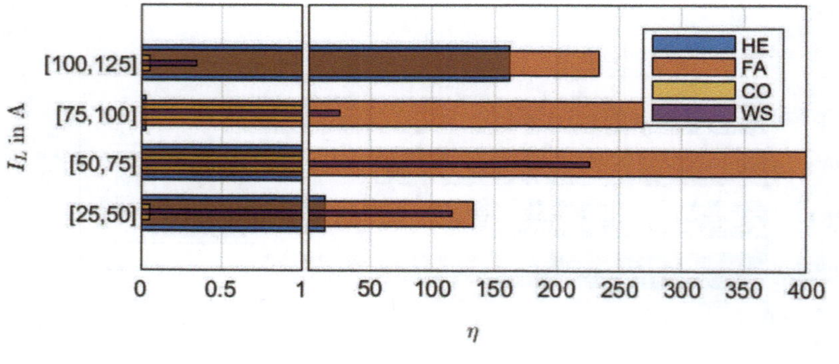

Figure 3.13: Deviation in η Caused by Different Faults at $N = 3000$ rpm

Initially, the values of η increase significantly up to 500 % due to flaking fault and for all I_L. This can be explained by the sensitivity of this parameter to rotational speed, where the flaking fault FA clearly reveals its effect at most. In addition to flaking, the wire string WS and corrosion fault CO have a remarkable influence on the value of η specially at I_L = (25 to 75) A.

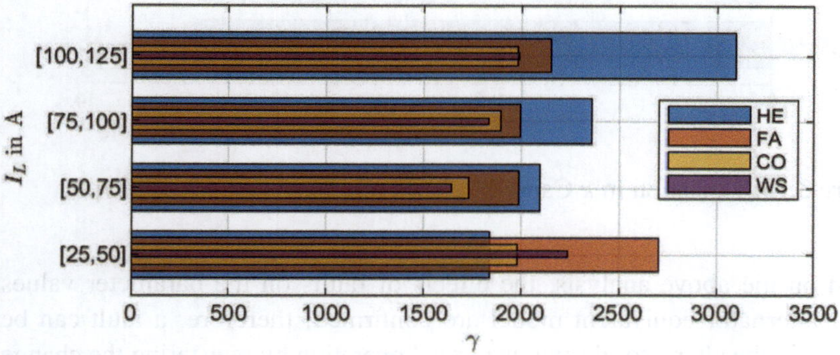

Figure 3.14: Deviation in γ Caused by Different Faults at N = 3000 rpm

With respect to γ, the presence of all types of faults has reduced its value between (5 to 35) % compared to the value of this parameter in the healthy machine. For lower electric loads between (25 to 50) A, the impact of FA is the largest and results in 50 % increase in γ. Followed by WS with up to 21 % and then CO up to 7 %. The deviations in γ are shown in Figure 3.14.

For the parameter κ, its value is reduced due to WS and FA faults by (6 to 30) % for all electric loads. The CO fault however, has revealed itself with an increase of 10 % in κ in the middle electric load range between (50 to 100) A. These deviations are shown in Figure 3.15.

Finally, the effect of all types of faults on λ, which is shown in Figure 3.16 is also significant resulting in a dramatic increase in its value for the range of electric load (50 to 125) A. The increase is remarkably caused by FA. However, for lower electric loads between (20 to 50) A, the values of λ register a significant decrease between 75 % and 100 % and for all types of faults.

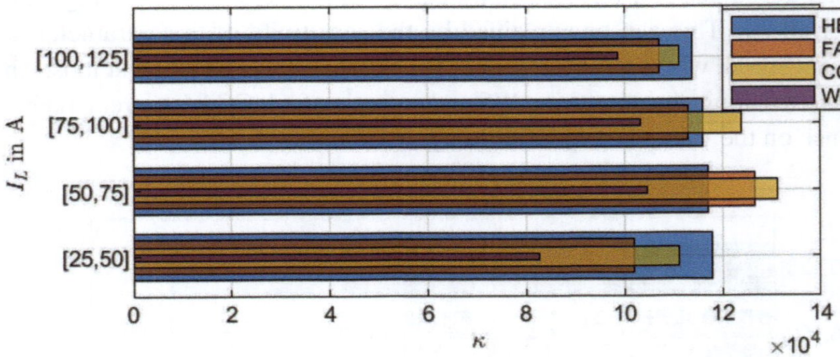

Figure 3.15: Deviation in κ Caused by Different Faults at N = 3000 rpm

Based on the above analysis, the effects of faults on the parameter values of the alternator equivalent model are confirmed; therefore, a fault can be detected in the alternator during its normal operation by registering the change in the parameter values of its simulation model. Further results for parameter deviations at different speeds are presented in Appendix B.

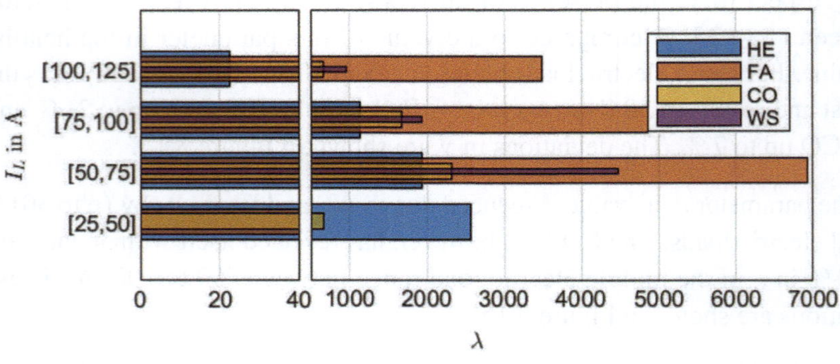

Figure 3.16: Deviation in λ Caused by Different Faults at N = 3000 rpm

3.2.3 Discussion and Analysis

Model-based fault detection based on parameter deviation of the equivalent simulation model proves to be effective. It successfully detects faults irrespective of their type.

The smallest detectable parameter change during these experiments in the selected operating range of the alternator is about 2 % of the respective value of the healthy machine. Therefore, a threshold of 2 % is established to determine when a parameter change is caused by faults. Additionally, this threshold value is necessary to exclude any tolerance caused by the solution of the optimization problem.

Moreover, it is proposed to implement the system test presented in Figure 3.10 on regular time intervals to exclude changes in parameter values due to aging and environmental effects.

Considering the analysis above, it is shown that each type of fault has an individual pattern affecting the parameter values of the alternator model. Furthermore, this pattern varies according to the current operating point as well as varies with respect to the individual parameter itself. In addition to that, the effect of faults is not necessarily an increase in parameter values, but rather a reduction has been equally registered. There is no doubt that fault detection based on parameter variation proves its effectiveness in this domain, nonetheless understanding the sequence of change of these parameters and correlating them to the type of fault or in other words performing fault identification is not trivial and requires machine learning classification algorithms. These algorithms reveal hidden correlations and sequences between parameter deviation and type of fault. Fault identification based on machine learning is presented in the following chapter.

4 Machine Learning-Based Fault Identification for Automotive Electric Machines

The previous Chapter 3 has demonstrated that the parameter values of the dynamic model are strongly dependent on the operating point of the alternator. Furthermore, a fault detection method has been presented based on investigating deviations in the values of model parameters due to the presence of faults. These deviations have some non-prominent patterns that are related to the type of fault causing them. Taking into consideration that these deviations also depend on the operating point, identifying faults based on parameter deviations is not a trivial task. A good candidate for this role is the ML-classification model. The former is capable of understanding hidden and complicated patterns between several observations and, accordingly, is capable of identifying the various faults based on parameter variations and the current operating point.

In this chapter, the key concepts of the ML-classification model for fault identification are introduced, providing a foundation to further extend the framework of model-based fault detection and identification for automotive electric machines. The work in this chapter extends the methods and concepts developed in [57].

This chapter is organized as follows: Section 4.1 introduces the ML model for fault identification including its structure, its inputs and outputs. Section 4.2 presents the results of the fault identification based on ML. Additionally, it provides a complete analysis of the ML model output prediction accuracy and comparison between several algorithms.

4.1 Machine Learning Model for Fault Identification

The framework for model-based fault detection that is presented in Subsection 3.2.1 and is illustrated in Figure 3.10 is further extended to include ML-based fault identification. Fault Identification (FI) in this context is defined as

A. Salah, *Model-Based Fault Diagnosis and Fault-Tolerant Control*,
Wissenschaftliche Reihe Fahrzeugtechnik Universität Stuttgart,
https://doi.org/10.1007/978-3-658-50108-2_4

the determination of fault type and location. The extended framework is shown in Figure 4.1.

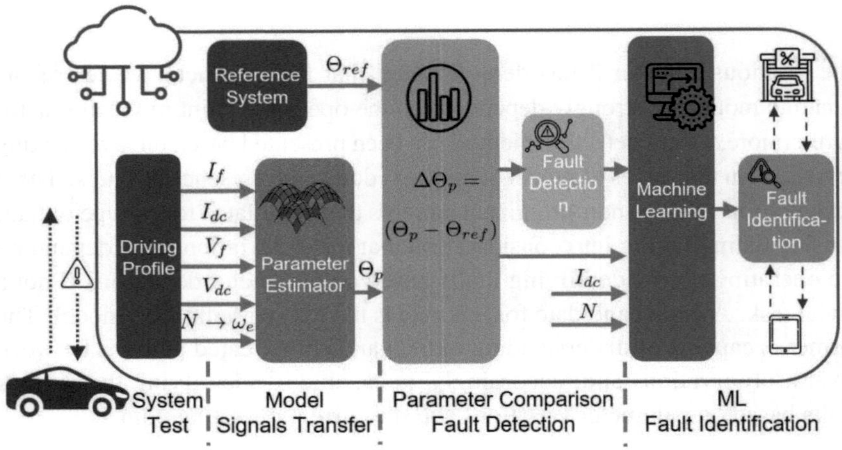

Figure 4.1: Framework of Model-Based Fault Detection and Identification for Automotive Electric Machines

Subsequent to Fault Detection (FD) based on the deviations of the parameters, this generates a signal that is in addition to the parameter deviation set Θ_p and the information about the current operating point in terms of $[N, I_{dc}]$ is transferred to a pre-trained ML model. The model performs supervised learning and based on that predicts the operating state of the alternator; whether it is healthy, with fault, and type of fault. Based on this, the block diagram representing this ML model is illustrated in Figure 4.2.

In general, the inputs of this ML model that are also defined as predictors, for training and testing, are divided into two groups:

- Operating point $[N, I_{dc}]$
- Mathematical model parameter deviations Θ_p.

It is worth mentioning that these inputs are either obtained from the available measurements in the vehicle or are generated within the cloud as a result of

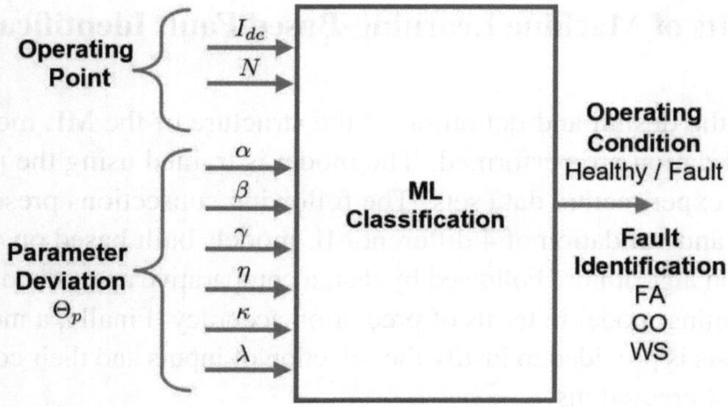

Figure 4.2: ML Model for Fault Identification

parameter estimation. For this ML-based fault identification there is no need to implement additional measurements or sensors.

The output is a prediction of the operating condition of the machine:

- Healthy HE.
- Fault and which type.

In this work, the following faults are considered for identification:

- Mechanical faults in the bearing of the drive side of the alternator:
  - Flaking Fault (FA).
  - Corrosion Fault (CO)
- Control signal fault of the voltage regulator caused by Wire String Fault (WS).

The ML model is trained using large data sets of the inputs and outputs mentioned above that are collected from the experimental setup mentioned in 2.2.1. The training is carried out using labeled data in order to implement supervised ML learning. Furthermore, the ML model is trained using several classification algorithms to investigate the best candidate algorithm in terms of accuracy and computational complexity.

4.2 Results of Machine Learning-Based Fault Identification

Following the design and definition of the structure of the ML model, training and validation are performed. The model is trained using the previously mentioned experimental data sets. The following subsections present results of training and validation of 4 different ML models built based on 4 different classification algorithms. Followed by that, a comparative analysis to define the best performing model in terms of prediction accuracy. Finally, a model sensitivity analysis is provided to justify the selection of inputs and their contribution to the output predictions.

4.2.1 Machine Learning Model Training, Validation and Testing

This Subsection presents results of Fault Identification (FI) based on machine learning. In addition, a detailed comparison of the best classification algorithms selected is provided highlighting their performance and applicability.

The performance of ML classifications models is best analyzed and understood with the help of a visualization method called; the confusion matrix. This visualization reveals how many times and in which class ML produced accurate predictions and which predictions were faults. The rows in this matrix are the true data, in this case they are the actual operating conditions of the alternator and the columns are the predicted states provided by the ML model. The diagonal elements of the matrix indicate the instances in which the predictions precisely match the actual or true values.

The complete data set is organized as follows:

- Training data: 90 % of the data set.
- Validation: Cross-validation of 5 partitions of the training data.
- Testing data:10 % of the data set that are not included for training.

After training, the ML algorithms with the highest validation accuracy are determined to be: Support Vector Machine (SVM), Wide Neural Networks (NN) and Quadratic Discriminant (QD) and their respective confusion matrices are illustrated in Figure 4.3. In addition to that, the confusion matrix of a model with poor performance is depicted in the same figure. This model is based on Naive Bayes (NB) algorithm.

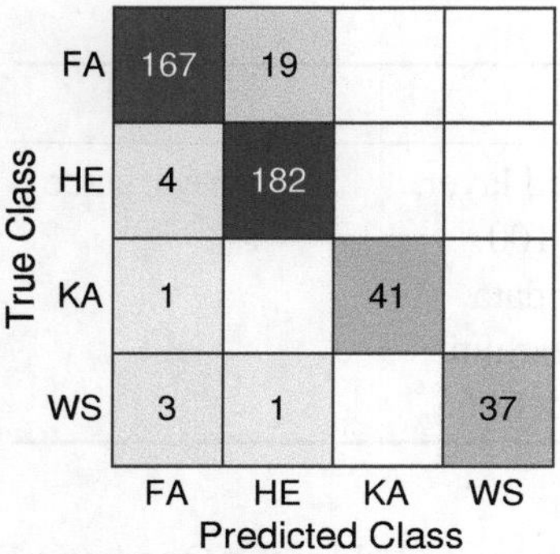

(a) Support Vector Machine (SVM)

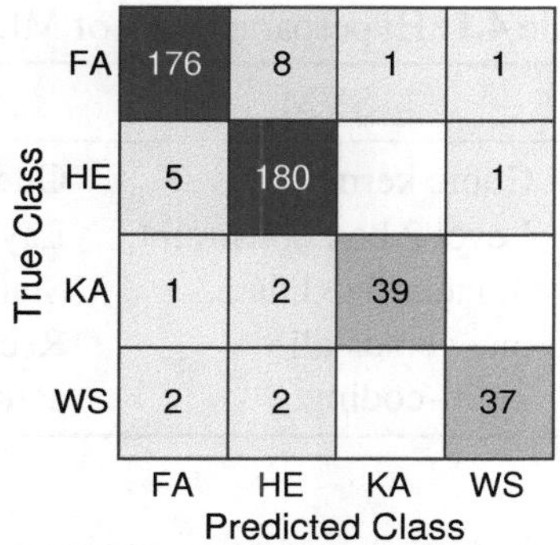

(b) Wide Neural Networks (NN)

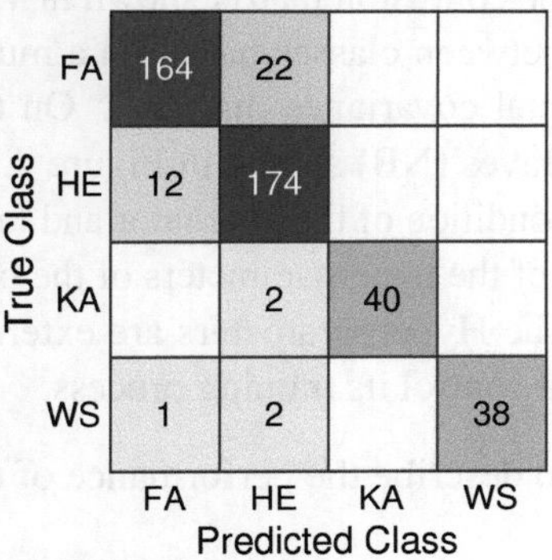

(c) Quadratic Discriminant (QD)

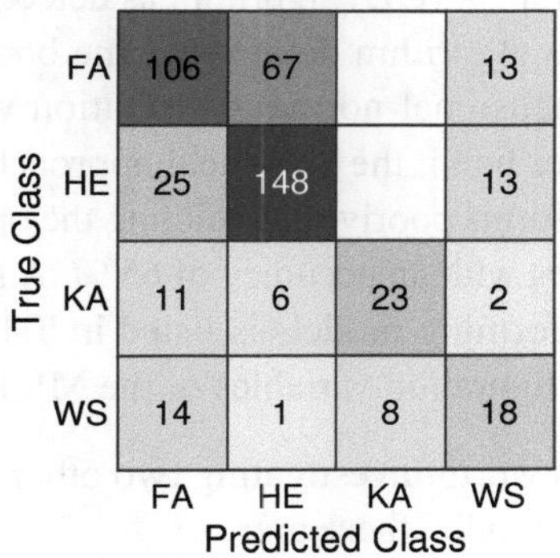

(d) Naive Bayes (NB)

Figure 4.3: Confusion Matrices of Validation Results

Both algorithms; SVM in Figure 4.3a and wide NN in Figure 4.3b have the highest prediction accuracy of 94 % and 95 %, respectively. For the predictions of SVM most of the fault predictions are with respect to HE and FA operating conditions. For the wide NN that are known to have a large number of neurons per layer, the same observations are made with respect to fault predictions between HE and FA with more scatter with respect the other operating states. In general, the prediction accuracy of both models SVM and wide NN is high;

therefore, they are considered successful and effective in identifying faults based on the deviations of the parameters.

Table 4.1: Hyperparameters of ML Models

SVM	NN	QD
Cubic kernel. Level 2 box constraint. Standardized data. one versus all- multi-coding.	One connected layer. Layer size is 100. Standardized data. Rectified linear unit activation function.	Full covariance

Another good prediction accuracy of 91 % is provided by the Quadratic Discriminant (QD) algorithm as detected from the confusion matrix shown in 4.3c. This algorithm determines the boundaries between classes based on a multi-dimensional normal distribution with unequal covariance matrices. On the other hand, the statistical approach Naive Bayes (NB) shown in Figure 4.3d, performs poorly in predicting the operating condition of the alternator and fault types with an accuracy of 65 %. A summary of the hyperparameters of the best performing models is listed in Table 4.1. The Hyperparameters are external configuration variables of the ML model that control its training process.

It is worth investigating two other metrics to describe the performance of the ML model, these are:

- True Positive Rate (TPR): the ratio of positive predictions that are correctly classified by the model to the total number of positive predictions according to Eq. 4.1:

$$TPR = \frac{TP}{(TP + FN)} \qquad \text{Eq. 4.1}$$

  where, TP is the number of true positive predictions and FN is the number of false negative predictions.

- False Negative Rate (FNR): is the ratio of positives that yields fault predictions according to Eq. 4.2:

$$FNR = \frac{FP}{(FP + TN)} \qquad \text{Eq. 4.2}$$

where, FP is the number of fault predictions and TN is the number of true negative predictions.

In general, the higher the value of TPR, the better is the accuracy of the trained model and the lower are the values of FNR. These metrics for the selected models are listed for each algorithm in Table 4.2 . As expected, both SVM and NN have the higher true positive rates, higher than 95 %. Remarkably, the relatively low value that is generated for the fault WS by the algorithm NN leaving the preference for SVM. On the other hand, the algorithm QD provides also relatively good rates above 90 %. With respect to NB as expected performs poorly according to these metrics as well with an average TPR of not more than 58 %.

Table 4.2: Comparison of TPR and FNR

Metric	FA	HE	CO	WS	Algorithm
TPR	90.3 %	97.8 %	97.6 %	90.2 %	SVM
FNR	9.7 %	2.2 %	2.4 %	9.8 %	
TPR	93.0 %	98.9 %	97.6 %	80.5 %	NN
FNR	7.0 %	1.1 %	2.4 %	19.5 %	
TPR	88.2 %	93.5 %	95.2 %	92.7 %	QD
FNR	11.8 %	6.5 %	4.8 %	7.3 %	
FNR	57.0 %	79.6 %	54.8 %	43.9 %	NB
FNR	43.0 %	20.4 %	45.2 %	56.1 %	

Based on the successful results of the validation, the ML models are required to be tested using new data that are not included in training. In this work, a random amount of 10 % of the prepared data set is used for testing. Figure 4.4 presents the confusion matrices produced by testing the selected ML models. It is essential to note that performing well in the validation phase and providing very good accuracies may not necessarily provide the same performance in the testing phase. Referring to the confusion matrices depicted in Figure 4.4, both models of SVM and wide NN have maintained their performance and accuracy with a slight reduction for SVM. In general, it is possible to improve the performance of testing by further tuning the hyperparameters of the respective algorithms. It is worth mentioning that over-fitting is avoided in this context by introducing the cross-validation.

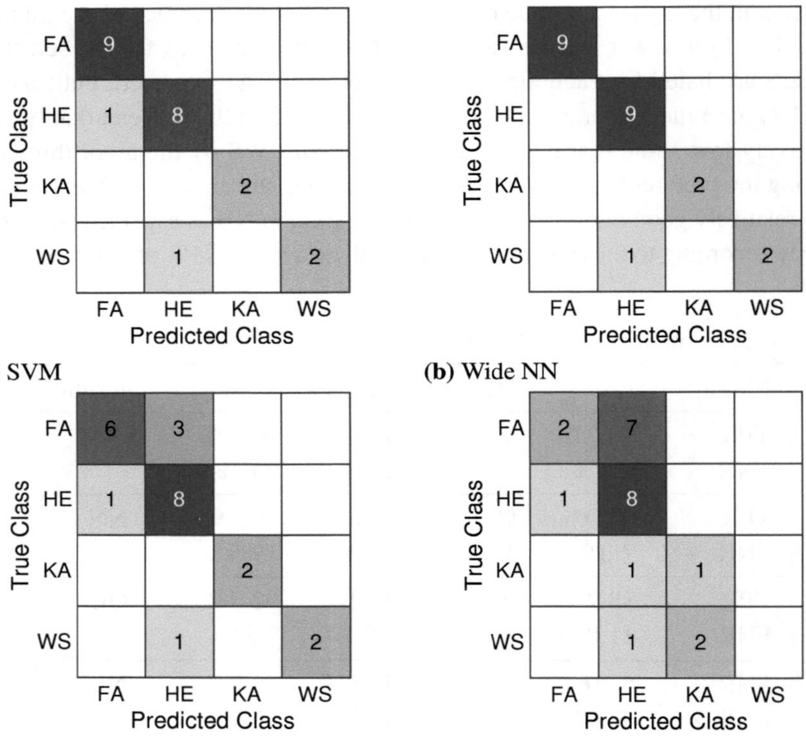

(a) SVM
(b) Wide NN
(c) QD
(d) NB

Figure 4.4: Confusion Matrices of Testing Results

In contrast, both models of QD and NB are substantially underperformed with a significantly decreased accuracies of 78 % and 48 % respectively, therefore, they are disqualified from the task of fault identification in automotive electric machines. This poor performance is depicted in the reduced concentration of predictions on the diagonal of the matrices in Figure 4.4c and Figure 4.4d. An overview of the accuracy of the selected models is given in Table 4.3.

Table 4.3: Prediction Accuracy of Validation and Testing

Accuracy	SVM	NN	QD	NB
Validation	94 %	95 %	91 %	65 %
Testing	91.3 %	95 %	78 %	48 %

4.2.2 Features Selection and Sensitivity Analysis

Sensitivity of the ML model to its inputs or predictors can be investigated according to several ranking criteria. The sensitivity analysis in this context defines the effect of these independent inputs on the output of the ML model. In other words, it determines which important predictor to include depending on different feature-ranking algorithms and the extent of this importance in comparison to others. Figure 4.5 presents the results of two ranking algorithms for inputs of the previously selected ML models. These ranking algorithms are; Anova [58] and Chi-Squared-Test [59]. The higher the score of the relevant input or feature, the higher its contribution or importance within the model.

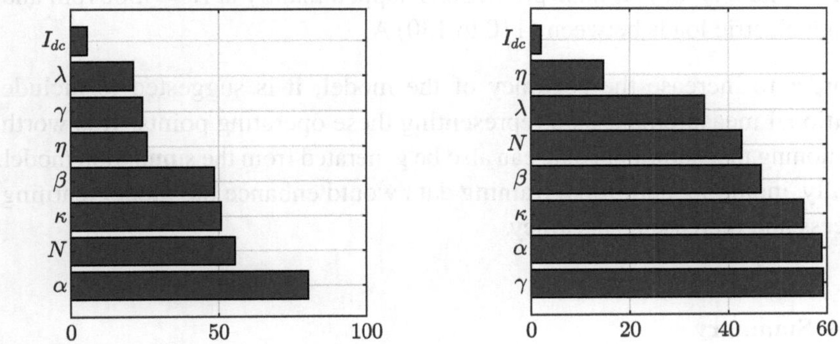

(a) Anova Algorithm **(b)** χ^2 Algorithm

Figure 4.5: Scores of the Feature Ranking Algorithms for Model Inputs

In Figure 4.5 it can be clearly seen that all selected features or inputs have relatively high scores and, consequently, a significant contribution to the model output. These scores are different with respect to each ranking algorithm. For example, the feature represented by the parameter α has the highest score of 60 % with respect to the Anova ranking algorithm. On the other hand, the highest scoring feature with respect to the χ^2 ranking algorithm is γ with a score of 60 %. In general, the order of the score is not completely relevant, but rather the values themselves should be high in order to conclude the relevance of the selected feature or input. One remarkable observation is that the input I_{dc} has relatively low scores in both ranking algorithms and therefore could be excluded to reduce the complexity of the model. A study of the resulting accuracy and performance of the ML model after removing the least relevant features is provided by [57].

4.2.3 Improving the Performance of the Machine Learning Model

Relatively increased false prediction rates are registered between the operating conditions HE and FA. Deeper analysis of these false predictions reveals that they occur primarily at specific operating points. For HE false predictions take place between (5000 to 5500) rpm at at medium loads I_{dc} is between (35 to 85) A. Whereas for FA, false predictions happen mainly at $N = 4500$ rpm and at high electric loads between (110 to 130) A.

In order to increase the accuracy of the model, it is suggested to include additional measurement data representing these operating points. It is worth mentioning that additional data can also be generated from the simulation model. Finally, including additional training data would enhance the model learning process and increase its accuracy.

4.2.4 Summary

The main objective of this chapter is to provide a novel approach to the problem of fault identification in automotive electric machines. In this context, ML models are trained and validated using experimental data. This approach provides an enhancement of the fault detection method presented in the previous chapter.

Prediction results and performance provided by trained ML models have shown a successful implementation. This success is evident in the high prediction accuracy of more than 90 % provided by both; a SVM and NN-based classification models. The second major contribution of this chapter is to provide a comparable study to evaluate the performance of several classification algorithms. In addition to that, a sensitivity analysis has been carried out in order to justify the selection of model inputs.

In order to improve the accuracy of the model, it is suggested to generate additional experimental data representing cases where an increased number of false predictions are generated. In addition to that, additional features can be generated from measurement data, such as time domain characteristics.

Further research work can be conducted to include the remaining useful life of the component and utilize it to predict the severity of the fault.

The next chapter introduces the final stage that follows Fault Detection and Identification (FDI), which is fault-tolerant control for automotive electric machines.

5 Fault-Tolerant Control Based on Model Reference Adaptive Control

The previous chapters have discussed fault detection and identification in automotive electric machines. It was demonstrated that these faults are not only frequent, but could also affect the performance of the vehicle with different levels and severity. The fault might be significant, potentially impacting the vehicle's functional safety, particularly when autonomous driving is taken into account. In this dissertation, the concept of Fault-Tolerant Control (FTC) is introduced to provide a solution to the degradation of the performance of automotive electric machines due to faults. This concept addresses the problem at an early stage, before the total breakdown of the vehicle; whether at the component level or the system level.

Fault-Tolerant Control (FTC) involves the design of special control concepts that are capable of accommodating faults in the actuators, the sensors or the overall system during operation, while maintaining an acceptable performance and ensuring stability. The main goal here is to mitigate the effect of faults on the overall performance and to retrieve to some extent the reference behavior. [38][60][23][61].

In the literature, FTC is categorized into two types based on their interaction with the presence of faults and whether the domain of action of the control concept extends beyond its execution level to the supervisory level within the system. The main types of FTC according to [62] are Passive Fault Tolerant Control (PFTC) and Active Fault Tolerant Control (AFTC).

In this dissertation, FTC for automotive electric machines is implemented based on MRAC as one type of AFTC. The classification of this technique among the others is shown in Figure 5.1 and is represented by the red line. MRAC belongs to Reconfiguration-Based Controls. In these techniques, controller parameters and structure are reconfigured automatically and online in response to a fault or deviation.

A. Salah, *Model-Based Fault Diagnosis and Fault-Tolerant Control*,
Wissenschaftliche Reihe Fahrzeugtechnik Universität Stuttgart,
https://doi.org/10.1007/978-3-658-50108-2_5

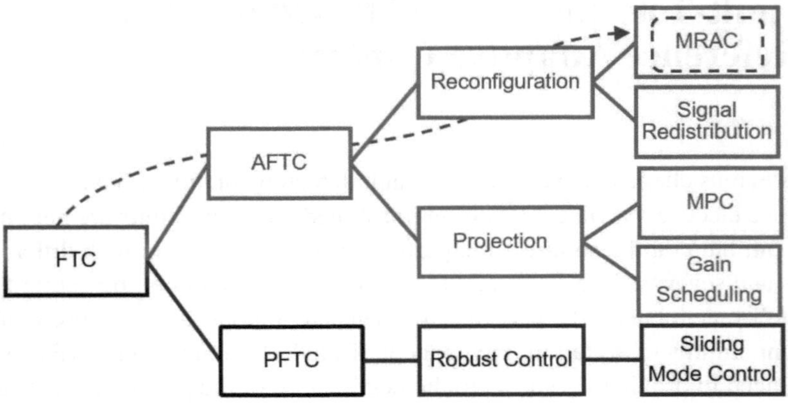

Figure 5.1: Different Methods of FTC According to its Classification

The main goal of this implementation is to maintain an acceptable performance of the electric machine, even in the event of mechanical, electrical or control signal faults. In addition to that, this implementation ensures the stability of the electric machine.

The advantages of implementing Model Reference Adaptive Control (MRAC) as a variant of AFTC, are outlined below.

- MRAC is basically designed on top of the existing controller of the machine, which implies that the main controller within the hardware is not necessarily changed. On the other hand, its parameters are tuned adaptively in response to faults or deviations from the response of the healthy or reference machine. Preserving the original controller structure which is called the nominal controller or the base-line controller is advantageous for existing serial products.
- It reacts to the presence of faults, deviations in the system and uncertainties, providing an advantage against PFTC techniques.
- MRAC does not require information about the presence of faults or deviations; however, the fault detection FD mechanism is integrated within the control law and the controller action takes place independently from the type of fault itself.

- The adaptation takes place online and is not pre-determined or accustomed to pre-determined faults or fault models, thus providing an advantage against projection-based control techniques.
- Lastly, the most important advantage of MRAC from the perspective of the control system is that global stability can be proved, its applicability in combination with different types of nominal controllers, and the ability to improve even the nominal behavior of the machine, especially in the presence of uncertainties.

This chapter is organized into two main parts; the first part presents the implementation of MRAC combined with an alternator that is controlled by a voltage regulator. Whereas the second part presents the implementation of MRAC combined with a traction motor that is controlled by Field Oriented Control (FOC). Section 5.1 provides a general introduction to the implementation of MRAC in automotive electric machine as a concept for FTC. Section 5.2 is dedicated to implementing MRAC in alternators and includes the related mathematical derivations, an introduction to the simulation model, and finally the simulation results are presented with the help of test cases combined with a detailed analysis. Similarly, Section 5.3 is dedicated to implementing MRAC for traction motors and includes the associated mathematical derivations, introduction of the simulation model, and finally the simulation results with the help of test cases are presented alongside an in-depth analysis.

5.1 Model Reference Adaptive Control for Automotive Electric Machines

In this dissertation, Model Reference Adaptive Control (MRAC) is implemented to handle the problem of fault-tolerant control in automotive electric machines. The proposed approach is a generalized concept and can be applied to alternators as well as to traction motors in battery electric vehicles or hybrid electric vehicles.

This type of adaptive control the MRAC, is applied to systems with uncertainties or unknown parameters and dynamics. The main structure of MRAC is depicted in Figure 5.2. This concept is generally applied on top of an existing **base-line controller** of the system. Principally, MRAC updates adaptively, the **parameters** or the **gains** of the base-line controller or the nominal controller which is already implemented in the system. **Parameter updates** are accomplished online during operation and based on an **adaptive control law**. An essential part of the MRAC control is the **reference model** which provides the desired **trajectory** of states in response to external or reference commands. At the same time, these commands are provided to the **actual system** so that the difference, which is defined as the **tracking error *e*** between the two trajectories ξ_{ref} and ξ_F, is registered and applied to the adaptive control law. The latter receives additionally, the **reference command** and the actual states then determines parameter adjustments. Finally, the adaptive control law forces the actual system output to follow the reference commands, which ensures an adequate closed-loop performance by minimizing the tracking error. In general, the reference model defines the base-line closed-loop performance.

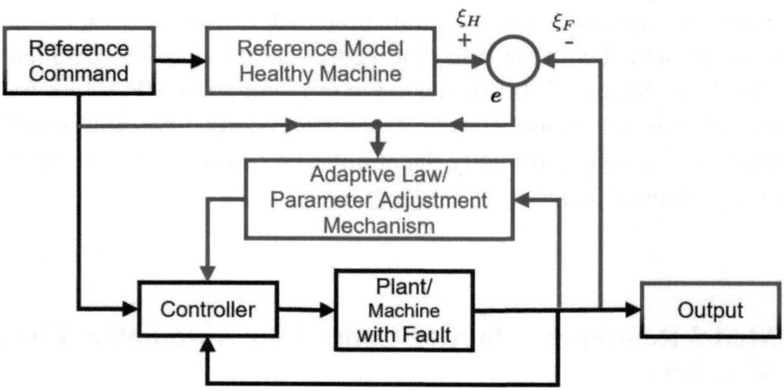

Figure 5.2: Main Structure of MRAC

The adaptive control law can be designed according to the following two methods:

- Based on a gradient approach.

- Based on stability theory, by introducing a Lyapunov candidate function. This approach is implemented in this dissertation because of its ability to ensure the closed-loop stability and prove it.

In this dissertation, analogies between model uncertainties and deviations of parameters resulting from faults are explored. These analogies are used to enhance the active implementation of MRAC in the context of FTC for automotive electric machines.

5.2 MRAC for Alternator Combined with PID Voltage Regulator

One of the main components in the automotive EPGS is the voltage regulator which, regulates the output voltage in other words the on-board voltage, monitors it and adjusts the excitation current I_f that is supplied to the rotor. Whenever the output voltage V_{dc} drops below a set level, the voltage increases the excitation current I_f, in order to increase the strength of the magnetic field and consequently the voltage. In contrast, if the voltage V_{dc} exceeds the set level, the excitation current I_f is reduced. This increase or decrease in excitation current is typically adjusted and controlled by the PWM voltage regulator. The following block diagram is construed to represent the mathematical model of the Electric Power Generation and Storage (EPGS) including the alternator and its control loop. The block diagram is shown in Figure 5.3:

Considering the common faults in the alternator; CO, FA to WS and DF or any fault provided by the battery, their effects are detectable within this control loop as illustrated in Figure 5.3. The following Subsection 5.2.1 presents the mathematical derivations of MRAC combined with the PID controller of the alternator. Then Subsection 5.2.2 presents the simulation model and the results of implementing the control configuration mentioned above. Finally, limitations of MRAC with respect to the alternator in addition to other aspects are discussed in Subsection 5.2.3.

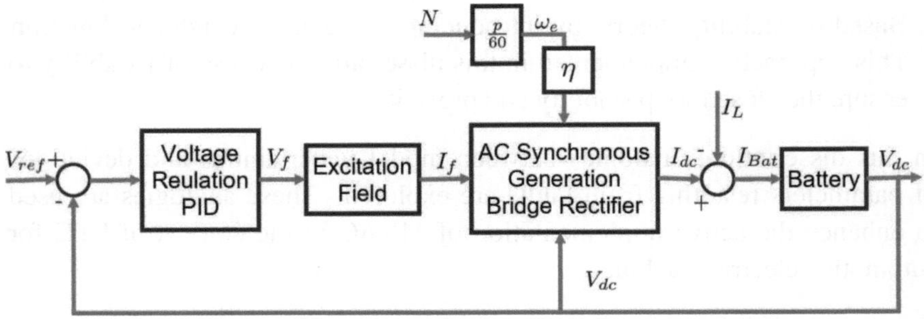

Figure 5.3: Functional Block Diagram of the Automotive EPGS Mathematical Model

5.2.1 Mathematical Derivations of MRAC Combined with the PID voltage regulator of the alternator

Recalling the simplified mathematical model of the alternator form Subsection 2.1.2 in the following equation:

$$\dot{I}_f = -\alpha I_f + \beta V_f \qquad \text{Eq. 5.1}$$
$$\dot{I}_{dc} = -\gamma I_{dc} + \eta\omega_e + \kappa I_f - \lambda V_{dc}$$

The Duty Cycle (DT) provided by the voltage regulator is represented by the equation Eq. 5.2:

$$DT = -K_p\left(V_{dc} - V_{ref}\right) - K_I \int_{t_0}^{t} \left(V_{dc} - V_{ref}\right).dt - K_d \dot{V}_{dc} \qquad \text{Eq. 5.2}$$

However the on-board voltage can be defined as in Eq. 5.3:

$$V_{dc} = I_{dc} R_o \qquad \text{Eq. 5.3}$$

where, R_o is the overall output impedance of the load and battery.

The excitation field voltage V_f is defined in terms of the duty cycle DT according to Eq. 5.1:

$$V_f = DT \cdot V_{dc} \qquad \text{Eq. 5.4}$$

Substituting Eq. 5.3 and Eq. 5.4 into Eq. 5.1, the linearized equivalent dynamic model of the alternator can be formulated according to Eq. 5.5:

$$\dot{I}_f = -\alpha I_f + \beta \cdot DT \cdot V_{dc}$$
$$\dot{I}_{dc} = -\gamma I_{dc} + \eta \omega_e + \kappa I_f - \lambda R_o I_{dc} \qquad \text{Eq. 5.5}$$

The state-space representation of the alternator model is defined as in Eq. 5.6:

$$\underbrace{\begin{bmatrix} \dot{I}_f \\ \dot{I}_{dc} \end{bmatrix}}_{\dot{X}} = \underbrace{\begin{bmatrix} -\alpha & 0 \\ \kappa & -(\gamma + \lambda R_o) \end{bmatrix}}_{A} \cdot \underbrace{\begin{bmatrix} I_f \\ I_{dc} \end{bmatrix}}_{X} + \underbrace{\begin{bmatrix} 1 & 0 \\ 0 & 1 \end{bmatrix}}_{B} \cdot \underbrace{\begin{bmatrix} \beta \cdot V_{dc} & 0 \\ 0 & \eta \end{bmatrix}}_{\Lambda} \cdot \underbrace{\begin{bmatrix} DT \\ \omega_e \end{bmatrix}}_{u} \qquad \text{Eq. 5.6}$$

$$y = \underbrace{\begin{bmatrix} 0 & R_o \end{bmatrix}}_{E} \cdot \underbrace{\begin{bmatrix} I_f \\ I_{dc} \end{bmatrix}}_{X}$$

where, $X \in R^2$ is the state vector, $A \in R^{2\times2}$ is the state matrix, $u \in R^2$ is the input vector, $B \in R^{2\times2}$ is the input matrix and $E \in R^{1\times2}$ is the output matrix.

Recalling the standard form for a linear system with structured and paramteric uncertainties from [63] according to Eq. 5.7:

$$X = AX + B\Lambda \left[u + \sigma(X)\right] \qquad \text{Eq. 5.7}$$

where, $\Lambda \in R^{2\times2}$ is the control effectiveness matrix and $\sigma(X)$ is the in-parameter state-dependent matched uncertainty and is defined according to Eq. 5.8:

$$\sigma(X) = W_o^T \cdot \Theta_o(X) \qquad \text{Eq. 5.8}$$

where, $W_o^T \in R^{2\times2}$ is the matrix of unknown parameters which are introduced by faults in the alternator and $\Theta_o(X) \in R^2$ is a vector containing the states or functions of the states.

The reference model $X_{ref} = X_H$, which defines the healthy state of the alternator is defined according to Eq. 5.9:

$$\dot{X}_H = A_H X_H + B\Lambda_H u_n \qquad \text{Eq. 5.9}$$

where, A_H and Λ_H include parameters of the healthy alternator. u_n is the base-line controller or the nominal controller that is designed according to the reference or the healthy machine.

Similarly, the model of the alternator-with-fault is defined in Eq. 5.10:

$$\dot{X} = A_F X + B\Lambda_F u_n \qquad \text{Eq. 5.10}$$

According to Subsection 3.2.2, the various faults in the alternator introduce deviations in parameter values that is represented by Eq. 5.11:

$$\Delta\Theta_p = \Theta_{p_F} - \Theta_{p_H} \qquad \text{Eq. 5.11}$$

Substituting Eq. 5.11 into Eq. 5.10, the model of the alternator with fault can be redefined according to Eq. 5.12:

$$\dot{X} = (A_H + \Delta A)\, X + B\,(\Lambda_H + \Delta\Lambda)\, u_n \qquad \text{Eq. 5.12}$$

Introducing MRAC in order to recover the reference behavior of the alternator and compensate effects of faults is achieved through a process comprising six steps, as outlined below.

Step1: Introducing the adaptive control signal as in Eq. 5.13:

$$u = u_n + u_a \qquad \text{Eq. 5.13}$$

where, u_a is the adaptive control signal which is provided by MRAC and is defined in **Step 2**. Substituting the overall control signal Eq. 5.13 into Eq. 5.12 the system with fault is defined according to Eq. 5.14:

$$\dot{X} = (A_H + \Delta A)\, X + B\,(\Lambda_H + \Delta\Lambda)\, u \qquad \text{Eq. 5.14}$$

Step 2: Re-formulation of the model in Eq. 5.14 and separation of terms into certain (Healthy) and uncertain (deviation due to faults):

$$\dot{X} = (A_H + \Delta A)\, X + B(\Lambda_F)\,(u_n + u_a) + B\Lambda_H \cdot u_n - B\Lambda_H \cdot u_n \qquad \text{Eq. 5.15}$$

$$\dot{X} = A_H X + B\Lambda_H u_n + B\Lambda_F\,(u_n + u_a) + \Delta A X - B\Lambda_H \cdot u_n$$

$$\dot{X} = A_H X + B\Lambda_H u_n + B\Lambda_F \left(u_n + u_a - \Lambda_F^{-1}\Lambda_H \cdot u_n\right) + \Delta A X$$

$$\dot{X} = \underbrace{A_H X + B\Lambda_H u_n}_{\text{Certain} \triangleq \text{Reference} \triangleq \text{Healthy}} \underbrace{+ B\Lambda_F \left(u_n + u_a - \Lambda_F^{-1}\Lambda_H \cdot u_n + (B\Lambda_F)^{-1}\,\Delta A X\right)}_{\text{Uncertain} \triangleq \text{System with Fault}} \qquad \text{Eq. 5.16}$$

In this step the term $(B\Lambda_H \cdot u_n)$ is added and subtracted to enable the separation of terms. The uncertain part of the model which is produced by the various types of faults in the alternator can be formulated according to Eq. 5.17:

$$W^T \cdot \Theta(X) = \underbrace{\begin{bmatrix} (B\Lambda_F)^{-1} \cdot \Delta A \\ \left(I - \Lambda_f^{-1}\Lambda_H\right) \end{bmatrix}}_{W^T} \cdot \underbrace{\begin{bmatrix} X \\ u_n \end{bmatrix}}_{\Theta(X)} \qquad \text{Eq. 5.17}$$

Although, the deviation in parameter values due to faults can be estimated according to Subsection 3.2.2, they are assumed in this context to be unknown in order to generalize the control action of MRAC and make it independent in compensating and mitigating effects of faults regardless of their type and severity.

Based on this, the adaptive control law u_a introduced by MRAC is designed in order to cancel the effect of uncertainties or parameter deviations and to augment the base-line controller. In order to achieve this purpose u_a is defined in Eq. 5.18:

$$u_a = -\hat{W}^T \cdot \Theta(X) \qquad \text{Eq. 5.18}$$

In other words, u_a provides an estimate of the actual uncertainty in the system and an augmentation of the base-line controller. Substituting Eq. 5.17 into Eq. 5.15 the model with fault can be defined in terms of the reference model according to Eq. 5.19:

$$\dot{X} = A_H X + B\Lambda_H \cdot u_n + B\Lambda_f \left(u_a + W^T \cdot \Theta(X)\right) \qquad \text{Eq. 5.19}$$

Step 3: Introduction of the nominal control or the base-line control u_n. For this alternator u_n has two parts; a PID controller that controls the duty cycle of V_f and a proportional gain controller for the electrical speed w_e. Based on that, u_n is defined according to Eq. 5.20:

$$u_n = \begin{bmatrix} -K_1(y - C_{v_{ref}}) - K_2 \underbrace{\int_0^t \left(y - C_{v_{ref}}\right) \cdot dt}_{Z} - K_3\dot{y} \\ \omega_e \end{bmatrix} \qquad \text{Eq. 5.20}$$

where, $\dot{Z} = y - C_{v_{ref}} = EX - C_{v_{ref}}$ and $y = EX$, Eq. 5.20 can be re-defined according to Eq. 5.21:

$$u_n = \begin{bmatrix} -K_1(EX - C_{v_{ref}}) - K_2 Z - K_3 E\dot{X} \\ \omega_e \end{bmatrix} \quad \text{Eq. 5.21}$$

Step 4 Defining the reference aggregated system based on substituting u_n from Eq. 5.21 into the reference model defined by Eq. 5.9:

$$\dot{X}_H = A_H X_H + B\Lambda_H \cdot \left[-K_1\left(EX - C_{v_{ref}}\right) - K_2 Z - K_3 E\dot{X}_H\right] \quad \text{Eq. 5.22}$$

Re-formulating Eq. 5.22:

$$\underbrace{\left(I + B\Lambda_H \begin{bmatrix} K_3 \\ 0 \end{bmatrix} E\right)}_{F} \dot{X} = \underbrace{\left(A_H - B\Lambda_H \begin{bmatrix} K_1 \\ 0 \end{bmatrix} E\right)}_{G} X - B\Lambda_H \begin{bmatrix} K_2 \\ 0 \end{bmatrix} Z + B\Lambda_H \begin{bmatrix} K_1 C_{v_{ref}} \\ \omega_e \end{bmatrix} \quad \text{Eq. 5.23}$$

$$\dot{Z} = EX - C_{v_{ref}}$$

Re-Writing Eq. 5.23:

$$F\dot{X} = GX - B\Lambda_H \begin{bmatrix} K_2 \\ 0 \end{bmatrix} Z + B\Lambda_H \begin{bmatrix} K_1 C_{v_{ref}} \\ \omega_e \end{bmatrix} \quad \text{Eq. 5.24}$$

$$\dot{Z} = EX - C_{v_{ref}}$$

Multiplying Eq. 5.24 by F^{-1}:

$$\dot{X} = F^{-1}GX - F^{-1}B\Lambda_H \begin{bmatrix} K_2 \\ 0 \end{bmatrix} Z + F^{-1}B\Lambda_H \begin{bmatrix} K_1 C_{v_{ref}} \\ \omega_e \end{bmatrix} \quad \text{Eq. 5.25}$$

$$\dot{Z} = EX - C_{v_{rev}}$$

The aggregated system of equation is defined according Eq. 5.26:

$$\xi_{ref} = \begin{bmatrix} X_H \\ Z_H \end{bmatrix} \quad \text{Eq. 5.26}$$

Re-defining Eq. 5.25 in terms of Eq. 5.26:

$$\dot{\xi}_{ref} = \underbrace{\begin{bmatrix} F^{-1}G & -F^{-1}B\Lambda_H \begin{bmatrix} K_2 \\ 0 \end{bmatrix} \\ E & 0 \end{bmatrix}}_{A_r} \underbrace{\begin{bmatrix} X_H \\ Z_H \end{bmatrix}}_{\xi_{ref}} + \underbrace{\begin{bmatrix} F^{-1}B\Lambda_H \begin{bmatrix} K_1 C_{v_{ref}} \\ \omega_e \end{bmatrix} \\ -C_{v_{ref}} \end{bmatrix}}_{B_r C} \quad \text{Eq. 5.27}$$

where, $A_r \in R^{3\times 3}$ is the aggregated reference state matrix and $B_r C \in R^3$ is the aggregated reference command matrix which contains the command values for the reference voltage and rotational speed.
The aggregated reference system in terms of A_r and $B_r C$ is in Eq. 5.28:

$$\dot{\xi}_{ref} = A_r \xi_{ref} + B_r C \quad \text{Eq. 5.28}$$

Step 5: Defining the aggregated system with fault by extending Eq. 5.27 in **Step 4** to include the uncertain term according to Eq. 5.19. The aggregated system-with-fault is defined Eq. 5.29:

$$\dot{\xi} = \underbrace{\begin{bmatrix} F^{-1}G & -F^{-1}B\Lambda_H \begin{bmatrix} K_2 \\ 0 \end{bmatrix} \\ E & 0 \end{bmatrix}}_{A_r} \underbrace{\begin{bmatrix} X_H \\ Z_H \end{bmatrix}}_{\xi} + \underbrace{\begin{bmatrix} F^{-1}B\Lambda_H \begin{bmatrix} K_1 C_{v_{ref}} \\ \omega_e \end{bmatrix} \\ -C_{v_{ref}} \end{bmatrix}}_{B_r C} \quad \text{Eq. 5.29}$$

$$+ \underbrace{\begin{bmatrix} F^{-1}B \\ 0 \end{bmatrix}}_{B_a} \Lambda_F \left(W^T \Theta (X) + u_a \right)$$

Similar to the concise representation in Eq. 5.28, the aggregated system model, which contains uncertainty or fault can be represented according to Eq. 5.30:

$$\dot{\xi} = A_r \xi + B_r C + B_a \Lambda_F \left(u_a + W^T \Theta(X) \right) \quad \text{Eq. 5.30}$$

Step 6 Defining dynamics of the tracking error e:

$$e \triangleq \xi - \xi_{ref} \quad \text{Eq. 5.31}$$

The dynamics of the tracking error is defined according to Eq. 5.32:

$$\dot{e} \triangleq \dot{\xi} - \dot{\xi}_{ref} \qquad \text{Eq. 5.32}$$

$$\dot{e} \triangleq A_r\left(\xi - \xi_{ref}\right) - B_a \Lambda_F \tilde{W}^T \Theta(X) \qquad \text{Eq. 5.33}$$

where $\tilde{W} = -\left(\hat{W} - W\right)$. The tracking error e defines the tracking dynamics of the system-with-fault. The adaptive update law is defined in Eq. 5.34:

$$\dot{\tilde{W}} = \gamma_a \Theta(X) e^T P_{Lyp} B_a \qquad \text{Eq. 5.34}$$

where, $\dot{\tilde{W}} = \dot{\hat{W}}$. From Eq. 5.32 and Eq. 5.17 into Eq. 5.34

$$\dot{\tilde{W}} = \gamma_a \begin{bmatrix} I_f \\ I_{dc} \\ u_n \end{bmatrix} \left(\xi - \xi_{ref}\right)^T P_{Lyp} B_a \qquad \text{Eq. 5.35}$$

where $\gamma_a > 0$ is the adaptive update rate or the learning rate is defined in Eq. 5.36:

$$\gamma_a = \begin{bmatrix} \gamma_f & 0 & 0 & 0 \\ 0 & \gamma_{dc} & 0 & 0 \\ 0 & 0 & \gamma_{DT} & 0 \\ 0 & 0 & 0 & \gamma_\omega \end{bmatrix} \qquad \text{Eq. 5.36}$$

and $P_{Lyp}$is the solution of the Lyapunov equation:

$$A_r P_{Lyp} + P_{Lyp} A_r^T + Q = 0 \qquad \text{Eq. 5.37}$$

where $Q > 0$.

The adaptive control signal in Eq. 5.18 can be re-defined according to Eq. 5.38:

$$u_a = \begin{bmatrix} DT_a \\ \omega_a \end{bmatrix} \qquad \text{Eq. 5.38}$$

where DT_a is the adaptive duty cycle and ω_a is the adaptive control signal of the speed. The proof of stability of MRAC and its control law are provided in Appendix C. The proof of stability and the related derivations are based on [63] and [48].

This implementation of MRAC and the utilization of the reference or healthy model of the alternator provide a good deployment of the digital twin concept.

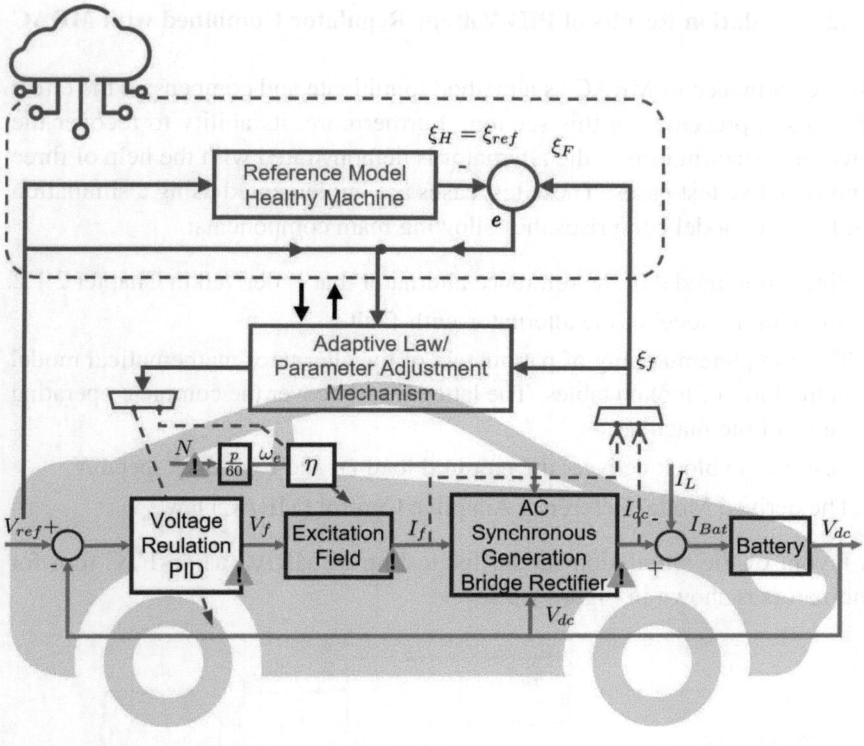

Figure 5.4: Block Diagram of MRAC for Alternator and DTw Deployment

With this configuration, the reference model of the alternator runs online and parallel to the real hardware in the cloud or a high- computational power server. Both machines exchange the necessary information about their operating states. Moreover, the adaptive control law that operates locally within the vehicle is implemented on top of the base-line controller. This implies that there is no need to replace the current controller supplied by the original manufacturer. In addition to that, the adaptive controller is not computationally expensive. A block diagram representing this control configuration is presented in Figure 5.4. The locations and types of faults that can be handled with this control configuration are also illustrated in the figure.

5.2.2 Simulation Results of PID Voltage Regulator Combined with MRAC

The performance of MRAC as a method to mitigate and compensate the effect of faults is presented in this section. Furthermore, its ability to recover the reference performance of the alternator is demonstrated with the help of three representative test cases. These test cases are implemented using a simulation model. The model comprises the following main components:

- Simulation model of the reference alternator that is derived in Chapter 2.1.2.
- Simulation model of the alternator-with-fault.
- The complete mapping of parameters of the alternator mathematical model in the form of lookup tables. The latter aims to cover the complete operating range of the machine.
- Test profile block delivers the required load I_L and rotational speed N.
- The derived Model Reference Adaptive Control (MRAC) law.

A layout of the simulation model for testing of MRAC as an FTC tool for alternators is shown in Figure 5.5.

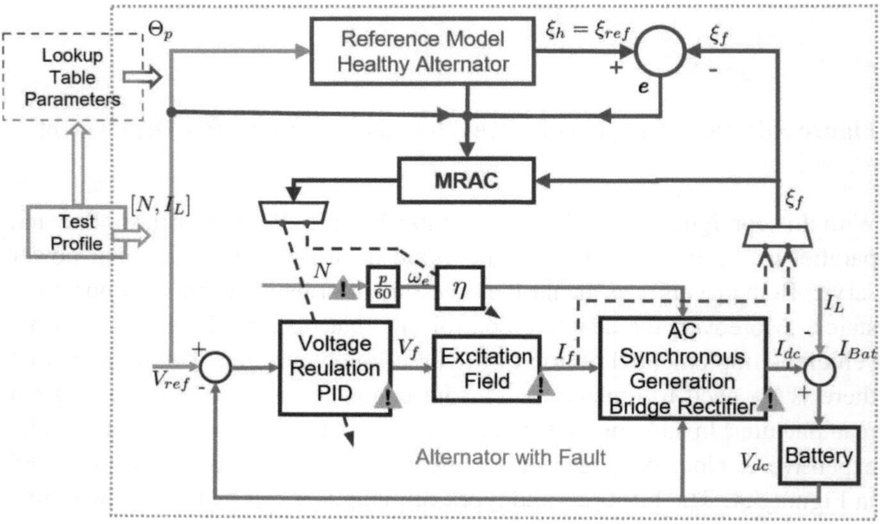

Figure 5.5: Layout of the Simulation Model of the Alternator with MRAC

Case 1: Fault with a Significant Effect on the Output of the Alternator

In this case, the fault has a significant effect on the output current of the alternator I_{dc} as well as on the excitation current I_f and is caused mainly by flaking and corrosion in the bearing of the drive side of the alternator. The test profile is designed to deliver a switching load current I_L in the range between (25 to 50) A with a frequency of 0.1 Hz and a constant rotational speed of N = 3000 rpm. This switching load is designed in such a way as to excite the system. It enables the adaptive controller to develop a proper adaptive learning mechanism, which in turn recovers the reference behavior of the alternator and mitigates the effect of faults. For this reason, the switching load is applied for about 8000 s. The test profile for **Case 1** is illustrated in Figure 5.6.

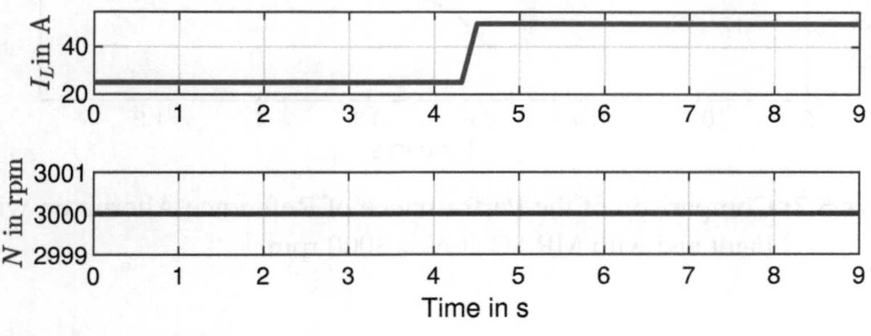

Figure 5.6: Test Profile for Case 1

As can be seen in Figure 5.7, the alternator-with-fault experiences a significant delay in delivering the required load current of 590 ms compared to the reference alternator. Moreover, this delay causes the on-board voltage of the vehicle to deviate significantly from the reference value of 28.5 V by −2 V which is harmful to the various electronics within the vehicle. In addition to that, the alternator-with-fault has to compensate for the required load current by increasing I_f up to 150 % percent compared to the reference one or by a factor of 3. The former results in a notable decrease in the efficiency of the machine due to increasing the excitation electric losses and increased mechanical losses.

In summary, it can be observed that the alternator voltage regulator with the help of the battery was able to achieve the required load, however, at the price of reducing the efficiency of the machine and a delayed response.

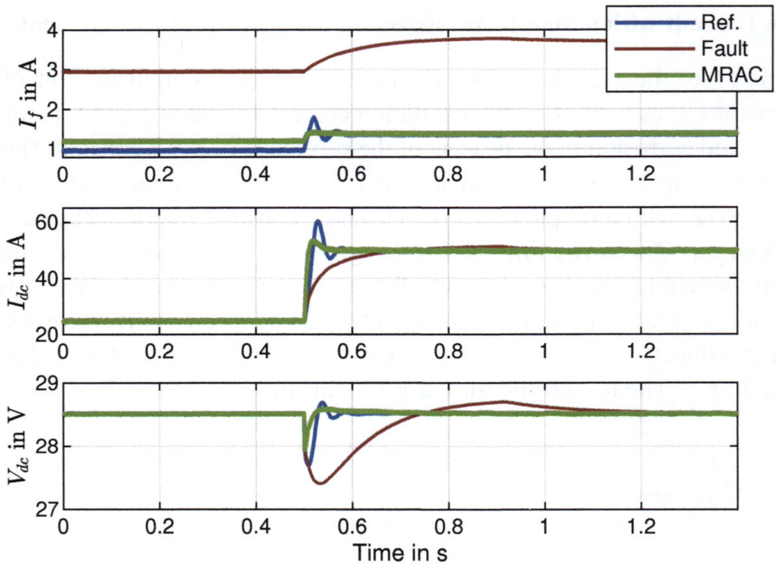

Figure 5.7: Comparison of the Performance of Reference Alternator, with Fault and with MRAC at $N = 3000$ rpm

On the other hand, and by implementing MRAC on top of the voltage regulator, it is possible to observe the significant improvement in the performance of the alternator-with-fault represented by the green line in the same Figure 5.7. With this implementation of MRAC on alternator-with-fault, the healthy performance is recovered by achieving the set point of the load current with minimum delay of less than 60 ms and the minimum tolerance of 0.6 % of the set value to 0.1 % of the reference system. The reference performance of the machine has not only been recovered, but also has been improved. In terms of the settling time, the alternator-with-fault and MRAC has achieved an improvement of 45 % compared to the reference system and an improvement of 70 % compared to the alternator-with-fault. In terms of overshoot, the alternator with MRAC has registered a remarkable improvement of 6 % overshoot compared to 20 % of the reference system where as the alternator-with-fault has performed much slower with a notable under-damped behavior.

The same observations can be concluded with respect to I_f and V_{dc}. However, it is possible to observe the reduced change in excitation ΔI_f to achieve the desired generator current between (25 to 50) A.

In addition to the alternator states I_{dc} and I_f, it is very convenient to investigate the evolution of the tracking error $e = \xi - \xi_{ref}$ over time during this test. The results of the tracking error are presented in Figure 5.8. The figure clearly illustrates how the tracking error between the reference alternator and the one with-fault and MRAC is decreasing and is asymptotically converging to zero as the test continues over time. This behavior of the tracking error indicates the good reference following behavior. Furthermore, it is essential to notice how the tracking error after 1000 s becomes bounded, indicating stability of the tracking performance.

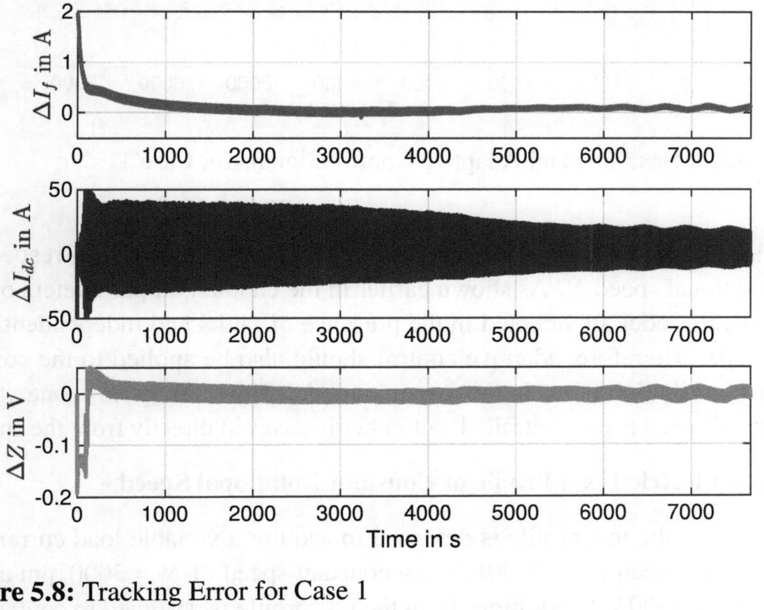

Figure 5.8: Tracking Error for Case 1

For this test, the adaptive control signals are shown in Figure 5.9. The upper part of the figure illustrates the *DT* delivered by the baseline-controller, namely the voltage regulator. The action of the MRAC controller is represented by the adaptively generated duty cycle u_{DT_a} that is illustrated in the center of Figure 5.9. Here, it can be seen how the signal evolves and interacts with the tracking error with the help of the adaptation rate γ_a, which allows it to counteract the effects of faults and the resulting uncertainties.

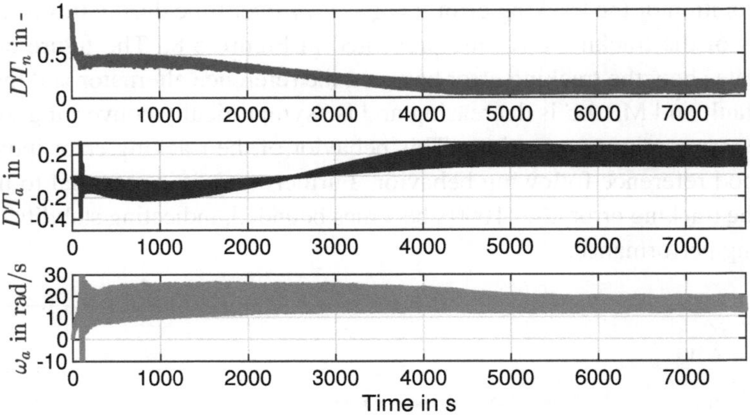

Figure 5.9: Base-line and Adaptive Control Signals for Case 1

The lower part of Figure 5.9 represents the adaptive signal with respect to the rotational speed N. As shown earlier in the chapter, all parameters of the equivalent model are affected in the presence of faults and independently of their type. Therefore, adaptive control should also be applied to the control of the contribution of ω_e in the dynamic model. However, in this context, the rotational speed is not controlled, but rather is received directly from the engine.

Case 2: Electric Load Profile at Constant Rotational Speed

In this case, the test profile is designed to acquire a variable load current I_L stepwise between $[30, 60, 90]$ A at a constant speed of $N = 3000$ rpm and a duration of 4500 s. In addition, I_L in the test profile is designed to contain an additional switching current of ∓ 3 A on top of each main step. This is necessary for the MRAC controller to achieve adequate adaptive and learning behavior. Switching of I_L enables the alternator to track the desired reference behavior by canceling out the undesired effects provided by the fault. The test profile for **Case 2** is provided in Figure 5.10.

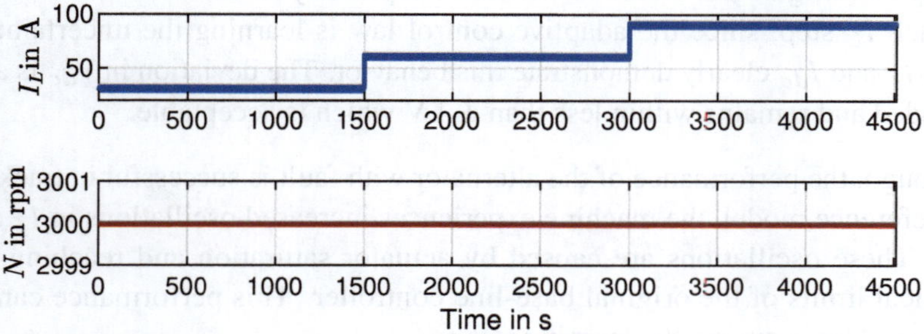

Figure 5.10: Test Profile for Case 2

The results of the implementation of the second test profile are shown in Figure 5.11.

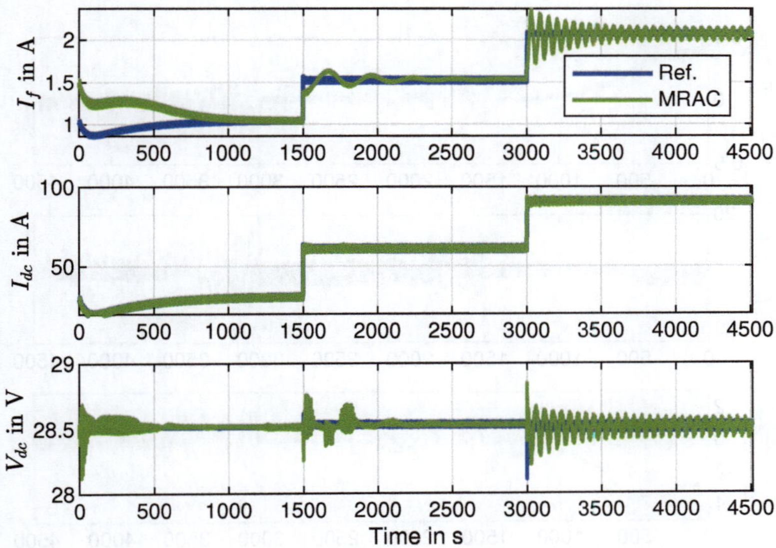

Figure 5.11: Performance of the Reference Alternator and Alternator-with-Fault MRAC for Test Case 2

Generally, the alternator-with-fault and MRAC is able to recover the reference behavior to a great extent. It can be clearly seen how the adaptive control is trying to mitigate the effect of the fault, especially in the first 500 seconds of each I_L step, since the adaptive control law is learning the uncertainties. Both I_f and I_{dc} clearly demonstrate this behavior. The deviation in V_{dc} is also bounded and remains within less than ± 1 V which is acceptable.

Although, the performance of the alternator with fault is successful in tracking the reference model, the machine experiences increased oscillations in I_f and I_{dc}. These oscillations are caused by actuator saturation and reaching the physical limits of the original base-line controller. This performance can be improved by re-tuning the learning rate γ_a

This successful performance in tracking the reference system is also evident from the behavior of the tracking error e which is shown in Figure 5.12. The former is asymptotically decreasing towards zero and is bounded. The maximum tracking error is 4.6 %,1.2 % and 0.03 % for e_{I_f}, $e_{I_{dc}}$ and $e_{v_{dc}}$ respectively.

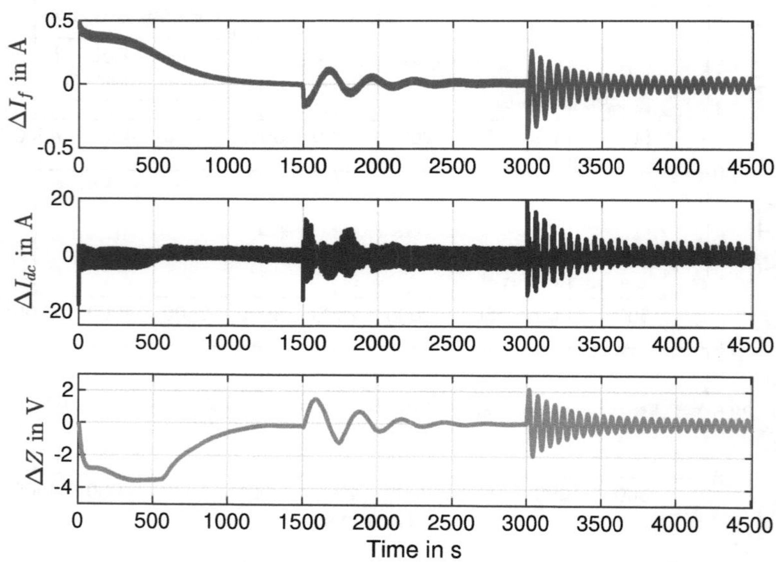

Figure 5.12: Tracking Error of Test Case 2

The control signals for this test case are presented in Figure 5.13. In the upper two parts of the figure, the control signals are represented by the base-line duty cycle and the adaptive one are presented. It is possible to notice the complementary behavior of these signals as they evolve throughout the test profile. Starting from the time instance of 3000 s, the duty cycle provided by the base-line controller reaches its maximum limit resulting in this oscillatory behavior. Moreover, as the base-line controller is designed by the manufacturer to operate with a specific frequency that is slower than the frequency of the adaptive controller, the complementary action of both controllers takes a relatively longer time to develop. Despite the oscillatory behavior, these oscillations are damped after 700 s of the change in I_L, at this time the effect of adaptation is sufficient to mitigate the effect of fault.

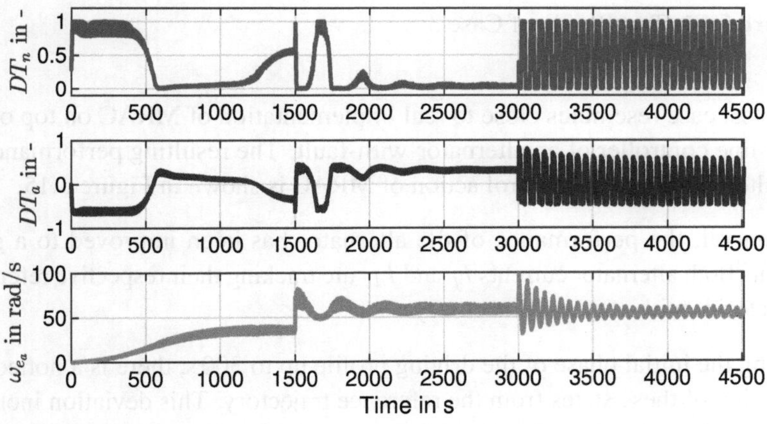

Figure 5.13: Base-line and Adaptive Control Signals for Case 2

Case 3: Representative Vehicle Driving Profile

The last test case implements a representative driving profile similar to the one provided by [52]. In this driving profile, both I_L and N vary as required by the vehicle and the driving road. The driving profile of this test is shown in Figure 5.14.

The round trip has a duration of 75 minutes and the alternator speed changes in the range between (3000 to 4000) rpm. The test vehicle requires a I_L in the range between (25 to 50) A during this trip.

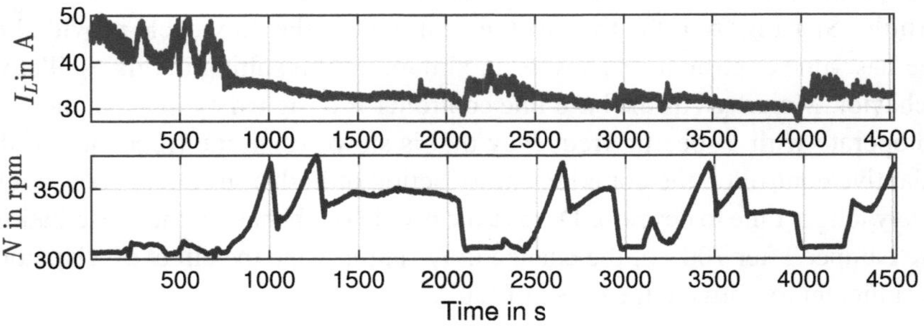

Figure 5.14: Test Profile of Case 3

This test case resembles close to real implementation of MRAC on top of the base-line controller of an alternator-with-fault. The resulting performance of the alternator with the control action of MRAC is shown in Figure 5.15.

In general, the performance of the alternator has been improved to a great extent. Both alternator currents I_f and I_{dc} are tracking their respective reference trajectories with very good accuracy.

During the initial phase of the driving profile up to 300 s, there is a noticeable deviation of these states from the reference trajectory. This deviation includes V_{dc}, which occurs during the learning process of the adaptation signal. Following the main learning phase, the controller achieves a good learning behavior and provides a very good compensation of fault effects.

The two deviation peaks in I_f at the time between (1000 to 1300) s and (2500 to 3500) s can be related to the nature of the driving profile. In these instances, there is a rapid and large increase in the speed. The learning rate of MRAC is not fast enough to cope completely and efficiently with this change, which implies the need to adjust the learning rates γ_a.

The average tracking error is 4.6 %,1.2 % and 0.03 % for e_{I_f}, $e_{I_{dc}}$ and $e_{v_{dc}}$ respectively, reflecting the successful performance of MRAC as illustrated in

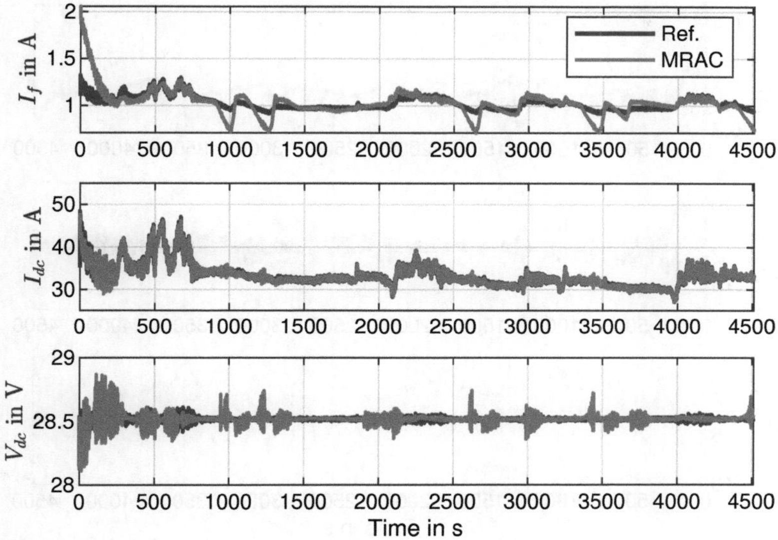

Figure 5.15: Comparison of the Performance of Reference Alternator, with Fault and MRAC for a Representative Driving Profile Case 3

Figure 5.16. This successful performance is evident by observing the minimum fluctuation in the on-board voltage.

Similarly in this case, the adaptation requires around 300 s in order to develop a good tracking behavior indicating a reduced tracking error compared to the one developed before this time.

By analyzing the control signals in Figure 5.17, the cooperative action of the two control signals is observable. Whenever DT of the base-line reaches its saturation limit, MRAC regulates it and provides the proper signal after canceling the effect of the fault. This phenomenon can be seen in several time instances and as an example in time duration between (300 to 700) s. The adaptive signal of speed w_{e_a} implies the need to increase the speed of the prime mover by a maximum of 30 rad/s.

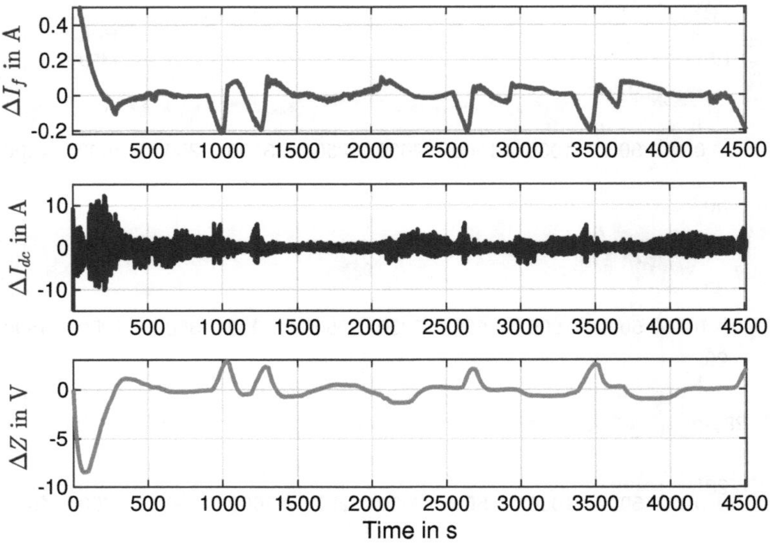

Figure 5.16: Tracking Error of Test Case 3

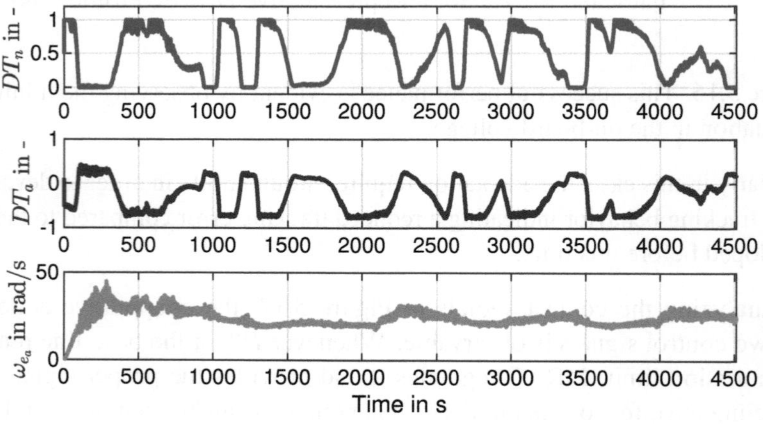

Figure 5.17: Base-line and Adaptive Control Signals for Case 3

5.2.3 Remarks and Limits of Performance Recovery

MRAC proved to have a successful enhancement to the PID controllers of the voltage regulator of alternators. It compensates effects of faults and recovers the reference performance to a great extent. The following aspects are emphasized about this implementation:

- In order to cover the entire operating range of the alternator, the learning or adaptation rate γ_a must be adaptively tuned. The reason for this issue arises from the structure of the alternator's mathematical model, which consists of lookup table containing the parameter mapping.
- Actuator saturation is an aspect that is critical for this implementation. Considering that the controlled signal is a duty cycle DT and it is the sum of both the nominal control signal u_n and the adaptive one u_a, in extreme cases of faults it is possible to reach the saturation limit of the actuator or the physical limitations of the base-line controller. A good candidate to address this problem is to implement what is known as barrier functions in addition to MRAC as proposed by [64] or performance recovery as suggested by [49].
- A critical aspect in designing MRAC is that its speed must be slower than that of the base-line controller. This is required to avoid canceling-out the course of action of the base-line controller. The two controllers are required to provide a consensus action. This is mainly the reason behind the extended time before a good learning behavior and therefore a good performance is developed. This aspect can be seen in **Case 1**.
- In this implementation with the alternator, MRAC provides a trade-off in minimizing the tracking error for I_f and I_{dc}. In this case, the adaptive controller guarantees tracking for the output state I_{dc} at the cost of increasing I_f.
- An individual draw back of this implementation is that the rotational speed is not provided as a controlled signal, and therefore the adaptive controller is not capable directly of influencing this input and it is up to the vehicle's engine to decide whether an increase or a change in speed is feasible.

5.3 MRAC for Traction Motor Combined with Field Oriented Control

Traction motors are automotive electric machines that are used to propel the vehicle. They are used in Battery Electric Vehicle (BEV) as the only source of torque and rotational speed. They transform the electrical energy received from the traction batteries into mechanical energy. In addition to that, they perform regenerative braking when the vehicle is decelerating and produce electrical energy that is used to charge the traction battery.

In the automotive domain, the most common traction electric motors are either synchronous; Permanent Magnet Synchronous Machine (PMSM), to EESM or asynchronous like squirrel cage machines. These types of motors are known for their high efficiency over a wide operating range and their ability to produce relatively high torques at low speeds. For their crucial role within BEV among other types of electrified vehicles, the traction motor needs to guarantee a fault-free operation and high performance.

In this work, fault-tolerant control based on MRAC is implemented on a Permanent Magnet Synchronous Machine (PMSM) in order to compensate the effect of common faults on the performance of the machine. The implementation is accomplished utilizing a simulation model introduced in [65]. The following Subsection 5.3.1 presents briefly the mathematical model of PMSM. After that, Subsection 5.3.2 introduces the derivations of MRAC combined with Field Oriented Control (FOC). Finally, the simulation model and the results of the implementation of the previously mentioned control system are presented in Subsection 5.3.3.

5.3.1 Mathematical Derivations of PMSM with FOC

The mathematical model of the PMSM is described within the magnetic flux reference frame according to Eq. 5.39 and is introduced by [66]:

$$V_d = I_d R_s + \frac{d\lambda_d}{dt} - \omega_e L_q I_q \qquad \text{Eq. 5.39}$$

$$V_q = I_q R_s + \frac{d\lambda_q}{dt} + \omega_e L_d I_d + \omega_e \lambda_{pm}$$

where,

$$\lambda_d = L_d I_d + \lambda_{pm} \qquad \text{Eq. 5.40}$$

$$\lambda_q = L_q I_q$$

and the torque equations:

$$T_e = \frac{3}{2} p \left(\lambda_{pm} I_q + \left(L_d - L_q\right) I_d I_q\right) \qquad \text{Eq. 5.41}$$

$$T_e = J \frac{d\omega_m}{dt} + B_f \omega_m$$

In order to generate the maximum torque, improve the performance of the traction motor and to operate it at higher speeds, a control technique called Field Oriented Control (FOC) is commonly implemented. It should be mentioned that, in this work FOC is introduced to explain the effective action of FTC in compensating the effect of faults in the traction motor. Details about the implementation and design of FOC are beyond the scope of this dissertation. The implementation presented in this work, supported by a simulation model, is meant to serve as a proof of concept and to generalize the proposed FTC based on MRAC for automotive electric machines.

An illustrative block diagram of this control technique is shown in Figure 5.18 and is based on [67]. The main goal of FOC is produce torque and maximize it to improve motor performance. This is accomplished when the stator magnetic field is orthogonal to the rotor magnetic field. The latter is achieved based on manipulating the three-phase currents in such a way that they result in this desired stator field vector according to the following main steps:

- The desired orientation of the stator field vector is determined on the basis of the current position of the rotor.

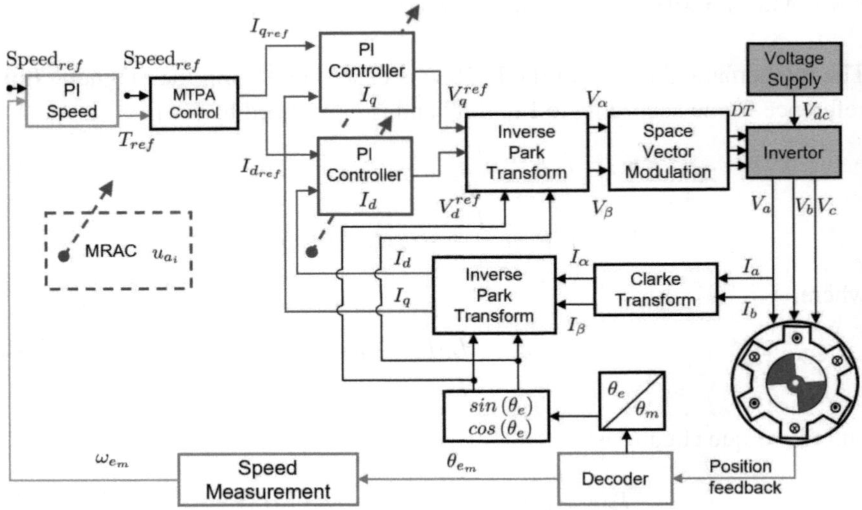

Figure 5.18: Block Diagram of FOC combined with MRAC for PMSM Traction Motor

- The space vector of the stator magnetic field needs to lead the reference space vector of the rotor by 90 degrees.
- The stator magnetic field is split into two components: the direct component I_d which is aligned with the rotor magnetic field and the quadrature component I_q which is orthogonal to the rotor magnetic field. Mathematically, this is called the Clarke/Park transformation. Here, the three-phase currents are transformed in the I_d and I_q and the stationary reference frame is transformed into a rotating reference frame.
- FOC forces the direct component I_d to zero and allows the quadrature component I_q to grow until the stator magnetic field aligns with the rotor field.
- The quadrature current I_q contributes to the torque while I_d does not produce any torque. Consequently, to generate the maximum torque, two PI-controllers are implemented to maximize I_q and set I_d to zero.

- The optimal reference currents $I_{d_{ref}}$ and $I_{q_{ref}}$ are determined based on Maximum Torque per Ampere (MTPA) and field weakening control techniques.
- Then, the PI-controllers output the voltages V_d and V_q.
- The above voltages are then converted to three-phase voltages using the inverse Park transform and space-vector modulation.
- The outer-control loop serves as a velocity controller and contains PI-controller to generate the reference torque T_{ref} based on the desired speed Speed_{ref}.

5.3.2 Field Oriented Control Combined with MRAC

Model Reference Adaptive Control (MRAC) is applied to the traction motor in order to compensate the effect of faults such as bearing damage, stator or rotor short and open windings to other faults including the control signal. This compensation is achieved by adjusting the parameters of the existing controllers adaptively. Traction motors are typically controlled using FOC concept which itself accommodates two PI-controllers that control the currents within the inner control loop (blue-line) in Figure 5.18. Typically, it is possible to control the desired speed based on the desired torque with the help of an additional PI-controller which creates the outer control loop represented by the (orange-line) in Figure 5.18. The adaptive controller is in this case applied only to inner control loop. The reason for this implementation is that the effect of faults and uncertainties are obvious within this control loop whether they are electrical or mechanical. In addition to that, and due to the inherent robustness of FOC, effects of faults can be rarely observed at the output in other words speed and torque. This issue will be discussed at the end of Subsection 5.3.3.

Recalling the generalized form of the system with structured and matched uncertainties:

$$\dot{\xi} = A_r\xi + B_rC + B_a\Lambda\left(W^T\Theta(X) + u_a + u_n\right) \qquad \text{Eq. 5.42}$$

The adaptive control signal u_a is defined as:

$$u_a = -\hat{W}^T\Theta(X) \qquad \text{Eq. 5.43}$$

where,

$$\dot{\hat{W}} = \gamma\Theta(X)\, e^T P_{Lyp} B_a \qquad \text{Eq. 5.44}$$

and e is the tracking error:

$$e = \xi - \xi_{ref} \qquad \text{Eq. 5.45}$$

and γ_a the learning rate is defined in this context according to Eq. 5.46:

$$\gamma_a = \begin{bmatrix} \gamma_d & 0 & 0 & 0 \\ 0 & \gamma_q & 0 & 0 \\ 0 & 0 & \gamma_{v_d} & 0 \\ 0 & 0 & 0 & \gamma_{v_q} \end{bmatrix} \qquad \text{Eq. 5.46}$$

The fault-tolerant controller is applied to the currents control loop or in other words the inner control loop, which is is represented by the blue-line in Figure 5.18. Within the inner control loop it possible to observe the effect of faults, whether they are mechanical or electrical and directly on the currents I_d and I_q and consequently T_e.

The aggregated state space representation of the **inner control loop** is defined according to Eq. 5.47:

$$\underbrace{\begin{bmatrix} \dot{I}_d \\ \dot{I}_q \end{bmatrix}}_{\dot{X}_i} = \underbrace{\begin{bmatrix} -\frac{R_s}{L_d} & \frac{\omega_e L_q}{L_d} \\ -\frac{\omega_e L_d}{L_q} & -\frac{R_s}{L_q} \end{bmatrix}}_{A_i} \underbrace{\begin{bmatrix} I_d \\ I_q \end{bmatrix}}_{X_i} + \underbrace{\begin{bmatrix} \frac{1}{L_d} & 0 \\ 0 & \frac{1}{L_q} \end{bmatrix}}_{B_i} \underbrace{\begin{bmatrix} V_d \\ V_q \end{bmatrix}}_{u_{n_i}} + \underbrace{\begin{bmatrix} 0 \\ -\frac{\omega_e \lambda_{mp}}{L_q} \end{bmatrix}}_{hc} \qquad \text{Eq. 5.47}$$

$$\begin{bmatrix} \dot{Z}_d \\ \dot{Z}_q \end{bmatrix} = \underbrace{\begin{bmatrix} 1 & 0 \\ 0 & 1 \end{bmatrix}}_{E_i} \begin{bmatrix} I_d \\ I_q \end{bmatrix} - \underbrace{\begin{bmatrix} I_{d_{ref}} \\ I_{q_{ref}} \end{bmatrix}}_{C_i} \qquad \text{Eq. 5.48}$$

The base-line controller u_{n_i} of this loop is according to Eq. 5.49:

$$\underbrace{\begin{bmatrix} u_{n_d} \\ u_{n_q} \end{bmatrix}}_{u_{n_i}} = -\underbrace{\begin{bmatrix} K_{1_d} & 0 \\ 0 & K_{1_q} \end{bmatrix}}_{K_{i_1}} \left(\begin{bmatrix} I_d \\ I_q \end{bmatrix} - \begin{bmatrix} I_{d_{ref}} \\ I_{q_{ref}} \end{bmatrix} \right) - \underbrace{\begin{bmatrix} K_{2_d} & 0 \\ 0 & K_{2_q} \end{bmatrix}}_{K_{i_2}} \begin{bmatrix} Z_d \\ Z_q \end{bmatrix} \qquad \text{Eq. 5.49}$$

The aggregated state space model representing the **inner control loop** for a traction motor containing uncertainties caused by faults is in Eq. 5.50:

$$\underbrace{\begin{bmatrix} \dot{X}_i \\ \dot{Z}_i \end{bmatrix}}_{\dot{\xi}_i} = \underbrace{\begin{bmatrix} A_i - K_{i_1} E_i & -K_{i_2} \\ E_i & 0 \end{bmatrix}}_{A_{ri}} \underbrace{\begin{bmatrix} X_i \\ Z_i \end{bmatrix}}_{\xi_i} + \underbrace{\begin{bmatrix} K_{i_1} \\ -I \end{bmatrix} C_i}_{BrC_i} + \underbrace{\begin{bmatrix} B_i \\ 0 \end{bmatrix}}_{B_{a_i}} \Lambda_{F_i} \left(u_{a_i} + W_i^T \Theta_i \right) \quad \text{Eq. 5.50}$$

and the adaptive control law for the currents u_{a_i} is:

$$u_{a_i} = -\hat{W}_i^T \Theta_i (X_i) \quad \text{Eq. 5.51}$$

where $\dot{\hat{W}}$ is defined in Eq. 5.52.

$$\dot{\hat{W}}_i = \gamma_i \underbrace{\begin{bmatrix} I_d \\ I_q \\ u_{n_d} \\ u_{n_q} \end{bmatrix}}_{\Theta_i (X_i)} \left(\xi_i - \xi_{i_{ref}} \right)^T P_{Lyp} B_{a_i} \quad \text{Eq. 5.52}$$

5.3.3 Simulation Results and Analysis of Field Oriented Control Combined with MRAC

For the implementation of MRAC as FTC technique for traction motors controlled by FOC, the simulation model provided in [65] is selected and extended. The simulation model implements a wide-speed operation of a PMSM drive. The drive uses a FOC system with maximum torque per ampere and field-weakening control techniques. The over-all simulation setup is shown in Figure 5.18. In this dissertation, two simulation models are added to [65] for each traction motor: the first represents a healthy motor and the second represents a motor-with-fault. Moreover, this injected fault is electrical as it is caused by damage in the stator windings causing inter-wire open circuit. Both machines operate in parallel and receive the same test profile. Additionally, a control system containing MRAC is implemented. Similarly to the alternator, the performance of FOC combined with MRAC is presented with the help of a test case.

Case 4: Speed Command with Variable Torque: In this test case, the traction motor is operated at its break-over speed and variable torques; between heavy load of 350 Nm and a light load of 50 Nm. The speed and torque values are selected so that the motor covers a wide operating range within the MTPA zone. Furthermore, a variable torque of ±20 Nm is applied to support the adaptive control law MRAC in developing a good learning behavior. The first 50 s segment of the test profile of **Case 4** is shown in Figure 5.19, however the overall duration is 300 s.

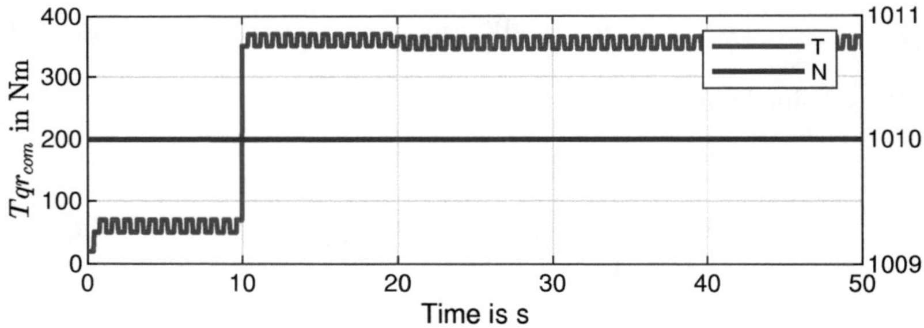

Figure 5.19: Test Profile of Case 4

The effect of the fault is significant on the controlled currents of the motor-with-fault. The direct current I_d reduces to zero with a delay of 0.5 s and an overshoot between −50 A and 25 A as illustrated in Figure 5.20. By implementing MRAC, it is possible to control this current to zero with reduced delays of less than 0.1 s and eliminated overshoots following by this behavior the reference motor. With this recovered behavior the traction motor is enabled to deliver the required output power.

A similar behavior is achieved by I_q, for the motor-with-fault. The reference current value of 350 A is achieved after a delay of 0.4 s compared to 0.08 s with MRAC. The overshoot in I_q is also significant for the motor-with-fault and is between(10 to 25) A. However, employing MRAC effectively eliminates these

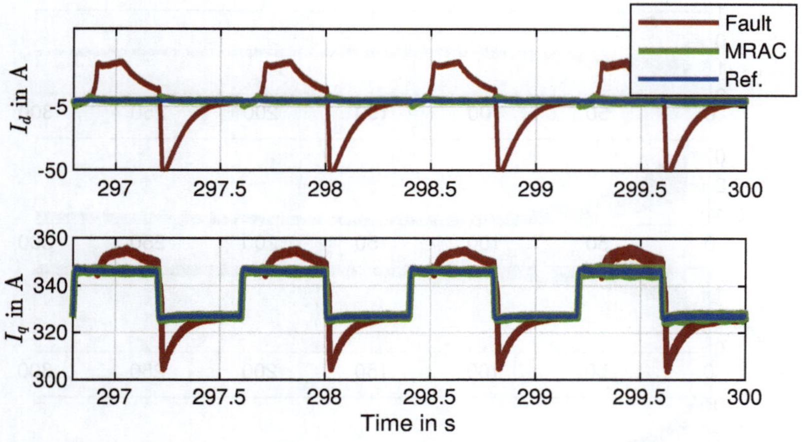

Figure 5.20: I_d and I_q for Reference, MRAC and Motor-with-Fault

overshoots, resulting in I_q behavior that closely resembles that of a healthy machine. This performance reflects the powerful course of action of MRAC.

The control signals of the motor with MRAC are shown in Figure 5.21. The control signals represent the per-unit-voltages of V_d and V_q. The figure illustrates how both the base-line controller and the adaptive MRAC are interacting with each other, until they result in a consensus and a cooperative control action to mitigate the effect of faults. The upper part of Figure 5.21 depicts the control signals of $\bar{V}_d$ while the lower part is for $\bar{V}_q$.

With respect to speed, the performance of the three machines is presented in Figure 5.22. The effect of fault is obvious on the performance of speed as it causes significant fluctuations by the motor-with-fault. These fluctuations are in the range of ±40 rpm by the end of the test. Compared to the healthy machine, the motor rotates smoothly with the applied load with minimum fluctuations of ±5 rpm.

The performance of torque shows a similar tendency to the speed, the motor-with-fault suffers from significant fluctuations between (250 to 400) Nm. It is worth mentioning here that the upper limit of torque provided by this motor is set to 400 Nm. Without this upper limit, the torque of the motor with fault may

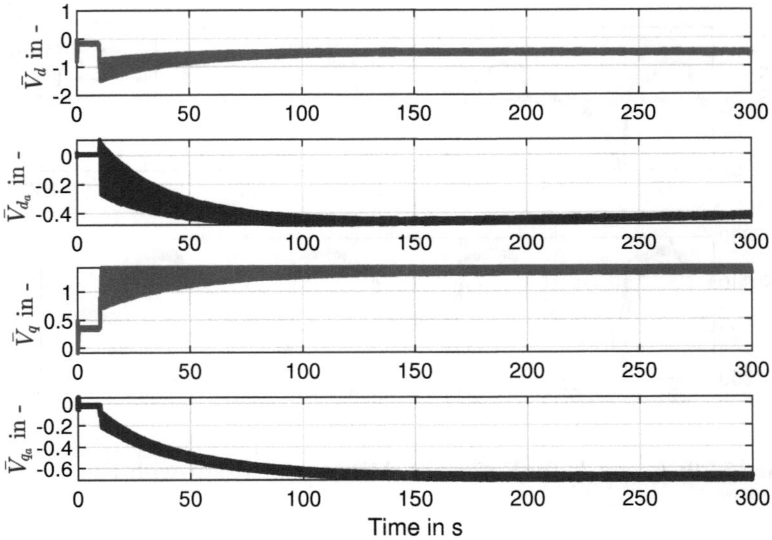

Figure 5.21: Adaptive and Base-Line Controller Signals for I_d and I_q

reach higher values. These fluctuations are reduced with the help of MRAC to be in the range between (340 to 380) Nm which is a deviation of 3 % with respect to the healthy motor. In addition to that, the torque remains below the saturation limits. The healthy motor operates normally between (345 to 365) Nm. This performance is recovered with the course of action of MRAC.

The output mechanical power is also affected by this fault. Figure 5.24 presents the electrical and mechanical power profiles of the three motors. The motor-with-fault is not able to deliver the required mechanical power accurately, as its power profile is deviating with 1.5 kW with respect to the one provided by the healthy motor. The motor with MRAC clearly provides the required mechanical power accurately. This shows in more ways the powerful performance of MRAC as the healthy motor tracking is successful with very good accuracy.

The implementation of MRAC allows for significant enhancement of speed performance, enabling the recovery of smooth rotational speed with fluctuation

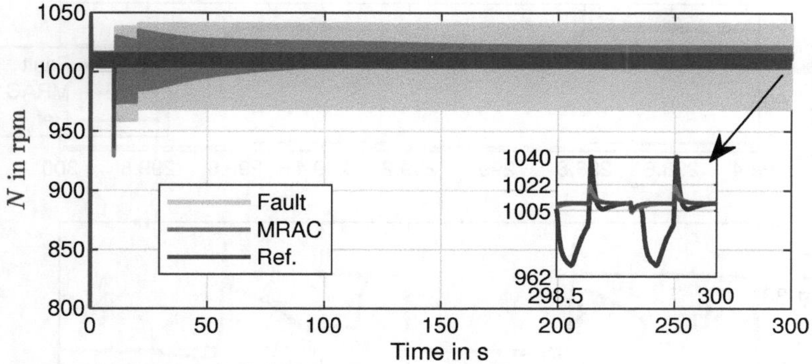

Figure 5.22: Speed Performance Case 4

minimized to ± 8 rpm. This recovery of speed performance is significant and is compatible with that provided by the healthy machine.

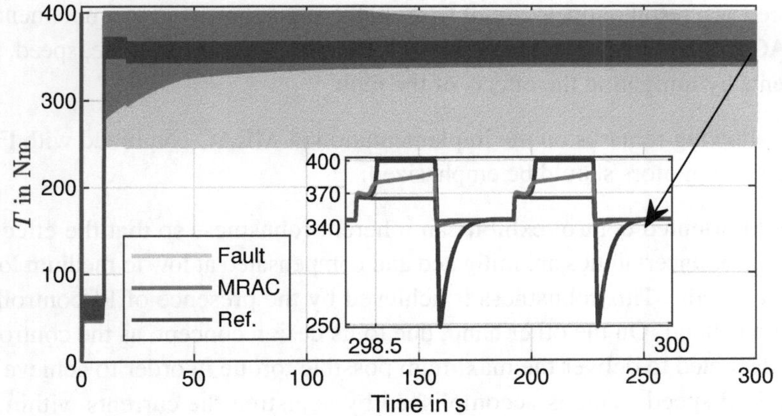

Figure 5.23: Torque Performance Case 4

With respect to the input electrical power, as the fault is caused by damaged stator windings, resulting in increased wire resistance and consequently lower inductance. Naturally, the electric power consumption will increase relative to the healthy motor. The electric power for both the motors with fault and the one with MRAC is 98 kW compared to 38 kW for the healthy motor. Based on

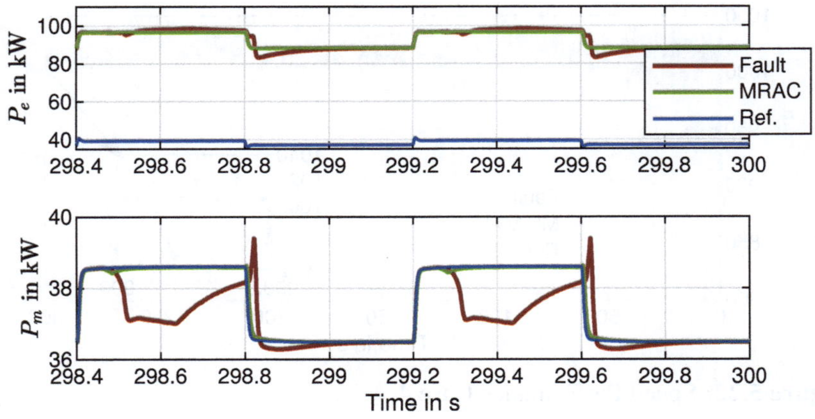

Figure 5.24: Power Profiles Case 4

that, it can be concluded that although the efficiency of the traction motor is reduced as a result of open-circuit wires in the stator, however, by implementing MRAC it is possible to maintain the performance in terms of torque, speed, and currents by mitigating the effects of the fault.

The following remarks on the implementation of MRAC combined with FOC for traction motors should be emphasized:

- Field oriented control exhibits an inherent robustness so that the effect of faults or uncertainties are mitigated and compensated at low to medium loads and speeds. This robustness is achieved by the presence of PI controllers on one hand. On the other hand, due to its design concept, as the controller is designed to deliver the maximum possible torque in order to achieve the desired speed. This is accomplished by adjusting the currents within the magnetic flux reference frame I_d and I_q.
- At high loads effects of faults and uncertainties on the performance become significant and FOC is no longer robust enough to recover the performance. In this case, the effectiveness of MRAC proves itself.
- The implementation of MRAC on the outer control loop is not effective. This control loop is basically designed to generate the reference torque based on

the desired speed. Effects of faults are rarely observed on the speed of a traction motor with this control configuration. This is due to the previously mentioned inherent robustness of FOC. The former issue can be explained by looking at Figure 5.25. This figure presents the torque, speed, and currents in the magnetic field reference frame for two motors; a healthy motor and a motor-with-fault. The fault is caused by high degree of corrosion in the drive-side bearing.

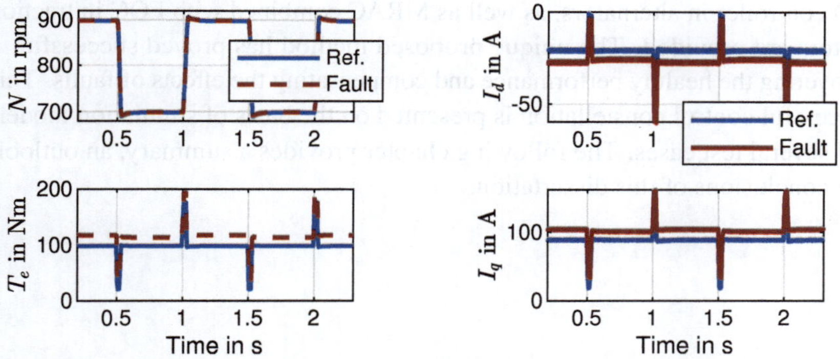

(a) Speed and Torque Profiles **(b)** Motor Currents Profiles

Figure 5.25: Comparison Between Reference Motor and Motor-with-Fault

On the left side at Figure 5.25a, both motors deliver the desired speed which is between (700 to 900) rpm. As the motor with fault experiences an increased friction due to corrosion, the electromechanical torque is increased by 14 % and accordingly the current I_d and I_d as shown in Figure 5.25b. This increase in T_e, I_d and I_q compensates the effect of corrosion fault at an earlier stage within the inner control loop so that the desired speed which is the output of the motor is not affected. For this reason, effects of faults within traction motors with the aforementioned control configuration, can only be addressed and compensated within the inner control loop.

- Finally, by designing MRAC, special attention has to be made in designing the speed of the adaptive control law. This speed is required to be slower than the speed of the inner control loop which is the base-line controller and at the same time fast enough to address effects of faults and compensate them. Also equally important, is to adjust this speed to be faster than the outer-loop controller to avoid interfering with its intended course of action. This speed is adjusted by selecting appropriate adaptive learning rates in order to achieve the above described trade-off.

This chapter has successfully presented the general concept of the adaptive control law MRAC as a method for fault-tolerant control for automotive electric machines. Detailed mathematical derivations of MRAC combined with the PID controller in alternators, as well as MRAC combined with FOC in traction motors are provided. The unique proposed method has proved successful in recovering the healthy performance and compensating the effects of faults. This successful control constellation is presented on the basis of simulation models and several test cases. The following chapter provides a summary, an outlook, and conclusions of this dissertation.

6 Conclusions and Outlook

The main objective of this work is to leverage the current on-board diagnostic process of the vehicle by introducing an advanced approach that elevates the current capabilities to a great extent. Although the current self-reporting of the diagnostic system is limited to simple and abstract fault codes, the key-components in the electrified and advanced powertrain require an advanced and reliable condition monitoring system. This system is required to facilitate continuous and direct fault detection and identification during vehicle operation.

The major contribution of this dissertation is to provide a novel multi-functional and model-based approach for fault detection and identification in automotive electric machines. This approach is designed according to a three-staged interconnected framework. Furthermore, it enables the vehicle to detect faults instantaneously and to address them directly long before the total breakdown of the components. Additionally, this approach utilizes the concept of digital twin to a great extent and parameter estimation as the core concept for fault detection. Subsequently, fault identification takes place in terms of type and location, which is provided by machine learning classification algorithms. The functionality of this approach extends to include fault-tolerant control using model reference adaptive control to mitigate the effects of faults and recover the reference performance of the electric machine, thus ensuring a good and safe driving experience.

6.1 Conclusions

- This work provided a simplified and linearized equivalent mathematical model for the electric machines. The alternator is selected as an example but the method is not limited to this implementation. This step is principle in preparation for the FD method. The equivalent model comprises parameters that provide an aggregated representation of several physical parameters.

A. Salah, *Model-Based Fault Diagnosis and Fault-Tolerant Control*,
Wissenschaftliche Reihe Fahrzeugtechnik Universität Stuttgart,
https://doi.org/10.1007/978-3-658-50108-2_6

Based on this representation, it is possible to distinguish the effect of several faults within the EPGS system of the vehicle. Additionally, this equivalent model has demonstrated high reliability and very good accuracy in simulating the machine under investigation according to the selected driving profiles.

- Parameters of the equivalent model are estimated utilizing experimental data and based on a global optimization problem. By implementing additional measures similar to starting from a large set of initial values, global minima and reproducibility of the solution are guaranteed. Based on that, this method proves its effectiveness in estimating parameter values of the equivalent model for both the healthy machine and the machine-with-fault.
- The final analysis of the results of this method have shown that the parameter values of the equivalent model are strongly dependent on the operating point of the machine. Therefore, comprehensive lookup tables are constructed, containing the complete mapping of parameter values within the selected operating range of the alternator.
- Comprehensive analysis of efficiency profiles of the alternators-with-fault have shown a significant decrease in the efficiency at a considerable range of output electric power and rotational speed. It is worth mentioning that this decrease in efficiency has a pattern that differs according to the type of fault. In addition to the efficiency, time domain characteristics of the alternator-with-fault are analyzed. The findings of this analysis have shown a significant change in the dynamic response of the alternator in terms of overshoot, settling time and steady-state error among others.
- Mathematical models are parameterized for alternators with different types of faults. The findings of the analysis have confirmed effects of faults on parameter values of the alternator model; therefore a fault can be detected in the alternator during its normal operation by registering the change or deviations in the values of parameter values with respect to the ones assigned to the model of the healthy machine.
- The analysis of parameter values of the various alternators has shown that each type of fault has an individual pattern in affecting the parameter value. Furthermore, this pattern varies according to the current operating point as well as varies with respect to the individual parameter itself. In addition to that, the effect of faults is not necessarily an increase in parameter values, but rather a reduction has been equally registered.

- Further investigations have demonstrated that, the threshold value for considering a parameter deviation as being caused by fault is 2 %. This threshold value is also necessary in order to exclude any tolerances caused by the solution of the optimization problem.
- Machine learning classification algorithms are implemented in this work to identify faults in terms of type and location. The motive behind this choice is, their effective capability to understand the hidden and complicated patterns between parameter deviations and the type of fault.
- Comparative study has been carried out about the performance of several ML classification algorithms in predicting the operating state of the alternator. This study has revealed that, the classification algorithms based on Support Vector Machine (SVM) and wide Neural Networks (NN) performed with the highest prediction accuracy exceeding 90 %. This high accuracy proves the successful implementation of ML classification for fault identification.
- This work extends the functionality of the novel approach to include Fault-Tolerant Control (FTC) which is considered the third and last stage within the framework. This is one of major contributions of this dissertation. Additionally, the implementation of FTC based on Model Reference Adaptive Control (MRAC) is one of the first attempts in the domain of electric machines that aims primarily to recover the reference performance and to mitigate the effect of faults within the machine itself during operation.
- The adaptive control law has demonstrated a successful addition to both; the PID voltage regulator in the alternator and the Field Oriented Control (FOC) in the traction motor. Moreover, this implementation reveals that MRAC is a generalized concept and can be implemented on different types of electric machines.
- The successful and powerful performance of MRAC as a method to compensate and mitigate the effect of faults has been demonstrated based on comprehensive test cases for both the alternator as well as the traction motor.
- One of the main attributes and advantages of MRAC is that it is implemented on top of existing controller, needless to change the structure and design of the existing controllers.
- In closing, this multi-functional model-based fault detection and identification in addition to fault-tolerant control for automotive electric machines, is a major contribution that enhances the functionality of the automotive

diagnostics. It is considered as a first step towards the preparation for the demands of the future vehicle.

6.2 Outlook

- Future work considers the derivation of equivalent and simplified models for additional types of automotive electric machines. In these models, the effects of faults can be clearly distinguished through variations in values of parameters.
- Additional common faults can be explored, and the respective mathematical model can be parametrized accordingly.
- In order to improve the accuracy of the ML model for fault identification, it is recommended to generate more experimental data that cover the operational range where false predictions are provided.
- It is interesting to explore additional features and inputs to the ML model such as time domain characteristics. These features can be directly derived from the available measurements.
- Further research work can be conducted to integrate the remaining useful life of components, allowing for the prediction of the severity or degree of the fault.
- Despite the effectiveness and powerful course of action of MRAC, actuator saturation is a crucial aspect that impairs this effectiveness. Future work investigates the implementation of barrier functions and performance recovery techniques.
- Further research may consider an adaptive tuning of the learning rates of MRAC. This aims to provide a faster and adequate learning behavior that covers the full operating range of the electric machine.

Bibliography

[1] Siwei Cheng, Yilu Zhang, Stephen Nawrocki, Thomas G. Habetler, and Mutasim Salman. Using only the dc current information to detect stator turn faults in automotive claw-pole alternators. In *8th IEEE Symposium on Diagnostics for Electrical Machines, Power Electronics & Drives*, pages 119–125. IEEE, 2011. ISBN 978-1-4244-9301-2. doi: 10.1109/DEMPED.2011.6063611.

[2] Antoine Picot, Ziad Obeid, Jeremi Regnier, Pascal Maussion, Sylvain Poignant, and Olivier Darnis. Bearing fault detection in synchronous machine based on the statistical analysis of stator current. In *IECON 2012*, pages 3862–3867, Piscataway, NJ, 2012. IEEE. ISBN 978-1-4673-2421-2. doi: 10.1109/IECON.2012.6389275.

[3] Gojko Joksimovic, Jaksa Riger, Thomas Wolbank, Nedjeljko Peric, and Mario Vasak. Stator line current spectrum content of a healthy cage rotor induction machine. In *8th IEEE Symposium on Diagnostics for Electrical Machines, Power Electronics & Drives*, pages 113–118. IEEE, 2011. ISBN 978-1-4244-9301-2. doi: 10.1109/DEMPED.2011.6063610.

[4] Bashir Mahdi Ebrahimi and Jawad Faiz. Feature extraction for short-circuit fault detection in permanent-magnet synchronous motors using stator-current monitoring. *IEEE Transactions on Power Electronics*, 25 (10):2673–2682, 2010. ISSN 0885-8993. doi: 10.1109/TPEL.2010.2050496.

[5] Jose Antonino-Daviu. Electrical monitoring under transient conditions: A new paradigm in electric motors predictive maintenance. *Applied Sciences*, 10(17):6137, 2020. doi: 10.3390/app10176137.

[6] Ji Pang, Weiguo Liu, Zhihuang Wei, Chenghao Sun, Ningfei Jiao, and Xu Han. Online diode fault detection in rotating rectifier of the brushless synchronous starter generator. *IEEE Transactions on Indus-

A. Salah, *Model-Based Fault Diagnosis and Fault-Tolerant Control*, Wissenschaftliche Reihe Fahrzeugtechnik Universität Stuttgart, https://doi.org/10.1007/978-3-658-50108-2

trial Informatics, 16(11):6943–6951, 2020. ISSN 1551-3203. doi: 10.1109/TII.2020.2974268.

[7] M. Q. Zhang and K. J. Tseng. An approach for on-line electrical machine winding inter-turn fault detection. In *2014 IEEE PES General Meeting | Conference & Exposition*, pages 1–5. IEEE, 27.07.2014 - 31.07.2014. ISBN 978-1-4799-6415-4. doi: 10.1109/PESGM.2014.6939379.

[8] Rongguang Hu, Jiabin Wang, Andrew R. Mills, Ellis Chong, and Zhigang Sun. Current-residual-based stator interturn fault detection in permanent magnet machines. *IEEE Transactions on Industrial Electronics*, 68(1): 59–69, 2021. ISSN 0278-0046. doi: 10.1109/TIE.2020.2965500.

[9] Oleg Gubarevych, Juraj Gerlici, Oleksandr Kravchenko, Inna Melkonova, and Olha Melnyk. Use of park's vector method for monitoring the rotor condition of an induction motor as a part of the built-in diagnostic system of electric drives of transport. *Energies*, 16(13):5109, 2023. doi: 10.3390/en16135109.

[10] Miguel Marques, João Martins, V. Fernão Pires, Rui Dias Jorge, and Luís Filipe Mendes. Fault detection and diagnosis in induction machines: A case study. In Luis M. Camarinha-Matos, Slavisa Tomic, and Paula Graca, editors, *Technological Innovation for the Internet of Things*, volume 394 of *IFIP Advances in Information and Communication Technology*, pages 279–286. Springer Berlin Heidelberg, Berlin/Heidelberg, 2013. ISBN 978-3-642-37290-2. doi: 10.1007/978-3-642-37291-9{\textunderscore}30.

[11] Oscar Duque-Perez, Luis-Angel Garcia-Escudero, Daniel Morinigo-Sotelo, Pedro-Esteban Gardel, and Marcelo Perez-Alonso. Analysis of fault signatures for the diagnosis of induction motors fed by voltage source inverters using anova and additive models. *Electric Power Systems Research*, 121:1–13, 2015. ISSN 03787796. doi: 10.1016/j.epsr.2014.11.021.

[12] Jun Hang, Yuanqi Li, Shichuan Ding, Chunyan Tang, and Qunjing Wang. *High-Resistance Connection Fault Severity Detection in a Permanent Magnet Synchronous Machine Drive System: 11-14 Aug. 2017.*

IEEE, Piscataway, NJ, 2017. ISBN 9781538632475. URL http://ieeexplore.ieee.org/servlet/opac?punumber=8049843.

[13] Puzakov Andrey Vladimirovich. Fault diagnosis in automotive alternators based on the output voltage parameters. In *2019 International Conference on Electrotechnical Complexes and Systems (ICOECS)*, pages 1–5. IEEE, 2019. ISBN 978-1-7281-1728-7. doi: 10.1109/ICOECS46375.2019.8949994.

[14] K. Gaeid K, H. Ping, M. Masood, and L. Szabo. Survey of wavelet fault diagnosis and tolerant of induction machines with case study. 2012.

[15] Mathworks. Rolling element bearing fault diagnosis - matlab & simulink - mathworks deutschland, 3/12/2020. Online Link.

[16] J. R. Stack, T. G. Habetler, and R. G. Harley. Fault classification and fault signature production for rolling element bearings in electric machines. pages 172–176, 2003. doi: 10.1109/DEMPED.2003.1234568.

[17] Jérôme Antoni. The spectral kurtosis: a useful tool for characterising non-stationary signals. *Mechanical Systems and Signal Processing*, 20(2): 282–307, 2006. ISSN 08883270. doi: 10.1016/j.ymssp.2004.09.001.

[18] Febus Reidj G. Cruz, Alejandro H. Ballado, Daniel Phillip Valmocena, Dwight G. Teves, and Jose Miguel Domingo. Vibrations signal analysis for generator machine fault detection. In *2020 4rd International Conference on Electrical, Telecommunication and Computer Engineering (ELTICOM)*, pages 205–209. IEEE, 03.09.2020 - 04.09.2020. ISBN 978-1-7281-8870-6. doi: 10.1109/ELTICOM50775.2020.9230497.

[19] R. Amiri and M. Mola. Inter-turn fault detection in pm synchronous motor by neuro-fuzzy technique. *International Journal of System Assurance Engineering and Management*, 11(5):923–934, 2020. ISSN 0975-6809. doi: 10.1007/s13198-020-01019-1.

[20] Bae Hyeon, Kim Sungshin, and Bae Jong-Il. Fault detection and diagnosis of winding short in electric machine based on park's vector. 2009. URL http://ieeexplore.ieee.org/servlet/opac?punumber=5256114.

[21] Raja Sekhar Kraleti, Maciej Zawodniok, and Sarangapani Jagannathan. Model based diagnostics and prognostics of three-phase induction motor for vapor compressor applications. In Suzanne Zheng, editor, *2012 IEEE Conference on Prognostics and Health Management (PHM 2012)*, pages 1–7, Piscataway, NJ, 2012. IEEE. ISBN 978-1-4673-0358-3. doi: 10.1109/ICPHM.2012.6299525.

[22] Rolf Isermann. *Fault-Diagnosis Systems: An Introduction from Fault Detection to Fault Tolerance.* Springer-Verlag Berlin Heidelberg, Berlin, Heidelberg, 2006. ISBN 9783540303688. doi: 10.1007/3-540-30368-5. URL http://site.ebrary.com/lib/alltitles/docDetail.action?docID=10229204.

[23] Halim Alwi, Christopher Edwards, and Tan Chee Pin. *Fault detection and fault-tolerant control using sliding modes.* Advances in Industrial Control. Springer, London, 2011. ISBN 9780857296498.

[24] X. Zhang, H. Uliyar, L. Farfan-Ramos, Y. Zhang, and M. Salman. Fault diagnosis of automotive electric power generation and storage systems. In *2010 IEEE International Conference on Control Applications*, pages 719–724, [Piscataway, N.J.], 2010. IEEE. ISBN 978-1-4244-5362-7. doi: 10.1109/CCA.2010.5611179.

[25] Annalisa Scacchioli, Giorgio Rizzoni, Mutasim A. Salman, Weiwu Li, Simona Onori, and Xiaodong Zhang. Model-based diagnosis of an automotive electric power generation and storage system. *IEEE Transactions on Systems, Man, and Cybernetics: Systems*, 44(1):72–85, 2014. ISSN 2168-2216. doi: 10.1109/TSMCC.2012.2235951.

[26] Alejandro J. Fernandez Gomez, Victor H. Jaramillo, and James R. Ottewill. Fault detection in electric motors by means of the extended kalman filter as disturbance estimator. In *2014 UKACC International Conference on Control (CONTROL)*, pages 432–437. IEEE, 2014. ISBN 978-1-4799-5011-9. doi: 10.1109/CONTROL.2014.6915179.

[27] Math Works. Fault detection using an extended kalman filter - matlab & simulink - mathworks deutschland, 13.08.2020.

URL `https://de.mathworks.com/help/predmaint/ug/Fault-Detection-Using-an-Extended-Kalman-Filter.html`.

[28] F. Karami, J. Poshtan, and M. Poshtan. Model-based fault detection in induction motors. In *2010 IEEE International Conference on Control Applications*, pages 1957–1962, [Piscataway, N.J.], 2010. IEEE. ISBN 978-1-4244-5362-7. doi: 10.1109/CCA.2010.5611214.

[29] S.S.S.R. Sarathbabu Duvvuri. Model-based bearing fault detection in induction motors under speed varying conditions. In *2018 8th IEEE India International Conference on Power Electronics (IICPE)*, pages 1–6. IEEE, 2018. ISBN 978-1-5386-4996-1. doi: 10.1109/IICPE.2018.8709523.

[30] Yilu Zhang, Satish Rajagopalan, and Mutasim Salman. A practical approach for belt slip detection in automotive electric power generation and storage system. In *2010 IEEE Aerospace conference proceedings*, pages 1–7, [Piscataway, N.J.], 2010. IEEE. ISBN 978-1-4244-3887-7. doi: 10.1109/AERO.2010.5446832.

[31] Moritz Benninger, Marcus Liebschner, and Christian Kreischer. Fault detection of induction motors with combined modeling- and machine-learning-based framework. *Energies*, 16(8):3429, 2023. doi: 10.3390/en16083429.

[32] G. Sajitha, N. Mayadevi, V. P. Mini, and R. Harikumar. Fault-tolerant control of bldc motor drive for electric vehicle applications. In *2019 International Conference on Power Electronics Applications and Technology in Present Energy Scenario (PETPES)*, pages 1–6. IEEE, 2019. ISBN 978-1-7281-2655-5. doi: 10.1109/PETPES47060.2019.9003925.

[33] Yongda Li and Pingping Gong. Fault-tolerant control of induction motor with current sensors based on dual-torque model. *Energies*, 16(8):3442, 2023. doi: 10.3390/en16083442.

[34] A. Fekih and F. N. Chowdhury. A fault tolerant control design for induction motors. In *2005 IEEE International Conference on Systems, Man and Cybernetics*, pages 1320–1325. IEEE, 2005. ISBN 0-7803-9298-1. doi: 10.1109/ICSMC.2005.1571330.

[35] D. Diallo, M.E.H. Benbouzid, and A. Makouf. A fault-tolerant control architecture for induction motor drives in automotive applications. *IEEE Transactions on Vehicular Technology*, 53(6):1847–1855, 2004. ISSN 0018-9545. doi: 10.1109/TVT.2004.833610.

[36] Mariem Ahmed Baba, Mohamed Naoui, and Mohamed Cherkaoui. Fault-tolerant control strategy for hall sensors in bldc motor drive for electric vehicle applications. *Sustainability*, 15(13):10430, 2023. doi: 10.3390/su151310430.

[37] Esmaeil Saffar, Mahmood Ghanbari, Reza Ebrahimi, and Mohammad Jannati. A simple fault-tolerant control method for open-phase three-phase induction motor drives. *Control Engineering Practice*, 136:105525, 2023. ISSN 09670661. doi: 10.1016/j.conengprac.2023.105525.

[38] Arslan Ahmed Amin and Khalid Mahmood Hasan. A review of fault tolerant control systems: Advancements and applications. *Measurement*, 143:58–68, 2019. ISSN 02632241. doi: 10.1016/j.measurement.2019.04.083.

[39] Zhiwei Gao, Carlo Cecati, and Steven Ding. A survey of fault diagnosis and fault-tolerant techniques part ii: Fault diagnosis with knowledge-based and hybrid/active approaches. *IEEE Transactions on Industrial Electronics*, page 1, 2015. ISSN 0278-0046. doi: 10.1109/TIE.2015.2419013.

[40] Jin Jiang and Xiang Yu. Fault-tolerant control systems: A comparative study between active and passive approaches. *Annual Reviews in Control*, 36(1):60–72, 2012. ISSN 13675788. doi: 10.1016/j.arcontrol.2012.03.005.

[41] Inseok Hwang, Sungwan Kim, Youdan Kim, and Chze Eng Seah. A survey of fault detection, isolation, and reconfiguration methods. *IEEE Transactions on Control Systems Technology*, 18(3):636–653, 2010. ISSN 1063-6536. doi: 10.1109/TCST.2009.2026285.

[42] Zexian Yang, Wei Zhang, Xingyan Liang, Jiale Wang, Liangguan Zhai, and Yufei Zhu. Position sensorless control of hybrid excitation axial

field flux-switching permanent magnet machine based on model reference adaptive system. In *2019 22nd International Conference on Electrical Machines and Systems (ICEMS)*, pages 1–5, Piscataway, NJ, 2019. IEEE. ISBN 978-1-7281-3398-0. doi: 10.1109/ICEMS.2019.8922248.

[43] Patarachai Chevathamanon, Ruili Wang, and Laszlo Farkas. Low-speed sensorless field-oriented control of asynchronous machine based on adaptive flux error suppression method. In Mihai Gavrilas, editor, *EPE 2016*, pages 205–208, Piscataway, NJ, 2016. IEEE. ISBN 978-1-5090-6129-7. doi: 10.1109/ICEPE.2016.7781333.

[44] Amir Bahador Javadi, Ali Akbar Damaki, Mostafa Shahnazari, and Ebrahim Amiri. Improved sensorless control of salient-pole permanent magnet synchronous motor based on fuzzy logic model reference adaptive system. In *2022 IEEE Kansas Power and Energy Conference (KPEC) 2022*, pages 1–4, Piscataway, NJ, 2022. IEEE. ISBN 978-1-6654-6591-5. doi: 10.1109/KPEC54747.2022.9814757.

[45] M. Vazifehdan, H. Abootorabi Zarchi, and M. Ayaz Khoshhava. Sensorless vector control of induction machines via sliding mode control based model reference adaptive system. In *2020 28th Iranian Conference on Electrical Engineering (ICEE)*, pages 1–6, Piscataway, NJ, 2020. IEEE. ISBN 978-1-7281-7296-5. doi: 10.1109/ICEE50131.2020.9260920.

[46] Mohammad Etemadi, Hamidreza Mosaddegh Hesar, and Mojtaba Ayaz Khoshhava. Speed control of brushless doubly fed induction machine drive based on model reference adaptive system. In Morteza Kheradmandi, editor, *2022 13th Power Electronics, Drive Systems, and Technologies Conference (PEDSTC)*, pages 305–309, Piscataway, NJ, 2022. IEEE. ISBN 978-1-6654-2043-3. doi: 10.1109/PEDSTC53976.2022.9767330.

[47] Md. Kamal Hossain and S. M. Rakiul Islam. Model reference adaptive control for surface permanent magnet synchronous machine. In *3rd International Conference on Electrical Information and Communication Technology*, pages 1–6, Piscataway, NJ, 2017. IEEE. ISBN 978-1-5386-2305-3. doi: 10.1109/EICT.2017.8275183.

[48] Tansel Yucelen. Model reference adaptive control. In John G. Webster, editor, *Wiley Encyclopedia of Electrical and Electronics Engineering*, pages 1–13. Wiley, 2000. ISBN 9780471346081. doi: 10.1002/047134608X.W1022.pub2.

[49] Tansel Yucelen and Eric Johnson. A new command governor architecture for transient response shaping. *International Journal of Adaptive Control and Signal Processing*, 27(12):1065–1085, 2013. ISSN 0890-6327. doi: 10.1002/acs.2376.

[50] V. Caliskan. Modeling and simulation of a claw-pole alternator: Detailed and averaged models, 2000.

[51] G. D. Marques. A simple and accurate system simulation of three-phase diode rectifiers. In *IECON '98. Proceedings of the 24th Annual Conference of the IEEE Industrial Electronics Society (Cat. No.98CH36200)*, pages 416–421. IEEE, 1998. ISBN 0-7803-4503-7. doi: 10.1109/IECON.1998.724276.

[52] Philip Rumbolz. *Untersuchung der Fahrereinflüsse auf den Energieverbrauch und die Potentiale von verbrauchsreduzierenden Verzögerungsassistenzfunktionen beim PKW*. Zugl.: Stuttgart, univ., diss., 2013, University of Stuttgart, Stuttgart, 2013.

[53] Rolf Isermann and Peter Ballé. Trends in the application of model based fault detection and diagnosis of technical processes. *IFAC Proceedings Volumes*, 29(1):6325–6336, 1996. ISSN 14746670. doi: 10.1016/S1474-6670(17)58696-6.

[54] M. Murken, D. Kubel, A. Kurz, A. Thanheiser, and P. Gratzfeld. Fault analysis of automotive claw pole alternator rectifier diodes. In *2018 IEEE International Conference on Electrical Systems for Aircraft, Railway, Ship Propulsion and Road Vehicles & International Transportation Electrification Conference (ESARS-ITEC)*, pages 1–6, Piscataway, NJ, 2018. IEEE. ISBN 978-1-5386-4192-7. doi: 10.1109/ESARS-ITEC.2018.8607641.

[55] Siddique Akbar, Toomas Vaimann, Bilal Asad, Ants Kallaste, Muhammad Usman Sardar, and Karolina Kudelina. State-of-the-art techniques for

fault diagnosis in electrical machines: Advancements and future directions. *Energies*, 16(17):6345, 2023. doi: 10.3390/en16176345.

[56] Alia Salah, Omar Abu Mohareb, and Hans-Christian Reuss. Model-based fault detection using parameter estimation in automotive epgs systems. In *2021 International Aegean Conference on Electrical Machines and Power Electronics (ACEMP) & 2021 International Conference on Optimization of Electrical and Electronic Equipment (OPTIM)*, pages 79–85. IEEE, 2021. ISBN 978-1-6654-0298-9. doi: 10.1109/OPTIM-ACEMP50812.2021.9590069.

[57] Alia Salah, Omar Abu Mohareb, and Hans-Christian Reuss. Fault diagnosis for automotive electric machines based on a combined machine learning and parameter estimation method: An approch for predective maintenance. In *2023 International Conference on Control, Automation and Diagnosis (ICCAD)*, pages 1–7. IEEE, 2023. ISBN 979-8-3503-4707-4. doi: 10.1109/ICCAD57653.2023.10152363.

[58] Yadolah Dodge. *The concise encyclopedia of statistics: With 247 tables*. Springer reference. Springer, New York, NY, 2008. ISBN 978-0-387-32833-1.

[59] Milton Abramowitz and Irene A. Stegun, editors. *Handbook of mathematical functions: With formulas, graphs and mathematical tables; [outgrowth of a Conference on Mathematical Tables, Cambridge, Mass., Sept. 15-16, 1954*. Dover books on intermediate and advanced mathematics. Dover Publ, New York, unabridged and unaltered republ. of the 1964 ed. edition, 1965. ISBN 978-0486612720.

[60] Youmin Zhang and Jin Jiang. Bibliographical review on reconfigurable fault-tolerant control systems. *Annual Reviews in Control*, 32(2):229–252, 2008. ISSN 13675788. doi: 10.1016/j.arcontrol.2008.03.008.

[61] John Baillieul and Tariq Samad, editors. *Encyclopedia of Systems and Control*. Springer eBook Collection. Springer, London, 2019. ISBN 978-1-4471-5102-9. doi: 10.1007/978-1-4471-5102-9.

[62] John Baillieul and Tariq Samad, editors. *Encyclopedia of Systems and Control*. Springer eBook Collection Engineering. Springer, London, 2015. ISBN 978-1-4471-5057-2. doi: 10.1007/978-1-4471-5058-9.

[63] Eugene Lavretsky and Kevin A. Wise. *Robust and adaptive control: With aerospace applications*. Advanced textbooks in control and signal processing. Springer, London and Heidelberg, 2013. ISBN 978-1-4471-4395-6.

[64] Poulomee Ghosh and Shubhendu Bhasin. State and input constrained model reference adaptive control, 2022. URL `http://arxiv.org/pdf/2206.13084`.

[65] Mathworks. Field-oriented control of an interior permanent magnet synchronous motor, 06.03.2025. URL `https://nl.mathworks.com/help/sps/ug/power-motordrive-IPMSM-FOC.html`.

[66] Mathworks. Mtpa control reference, 06.03.2025. URL `https://nl.mathworks.com/help/mcb/ref/mtpacontrolreference.html#mw_3ac2d1c8-b2be-4cf1-a867-5edd1d37c8f6`.

[67] Mathworks. Field-oriented control (foc), 05.03.2025. Online Link.

Appendix

A. Simplification of the Alternator Mathematical Model

The rotor-stator dynamics:

$$\dot{\mathbf{I}} = -\mathbf{L}^{-1}\left(\mathbf{RI} + \mathbf{V}\right) + \mathbf{L}^{-1}\left(\frac{d}{d\theta_e}\mathbf{L_f}\left(\theta_e\right)\omega_e I_f + \mathbf{L_f}\left(\theta_e\right)\dot{I}_f\right) \quad \text{A.1}$$

The dynamics of the field:

$$\dot{I}_f = -{L_f}^{-1}\left(R_f I_f + V_f\right) + {L_f}^{-1}\left(d/(d\theta_e)\, L_f\left(\theta_e\right)\omega_e I + L_f\left(\theta_e\right)\dot{I}\right) \quad \text{A.2}$$

Rectifier Bridge Switching function:

$$f = pg \quad \text{A.3}$$

$$f = \begin{bmatrix} f_a \\ f_b \\ f_c \end{bmatrix} \qquad p = \begin{bmatrix} \frac{2}{3} & -\frac{1}{3} & -\frac{1}{3} \\ -\frac{1}{3} & \frac{2}{3} & -\frac{1}{3} \\ -\frac{1}{3} & -\frac{1}{3} & \frac{2}{3} \end{bmatrix} \qquad g = \begin{bmatrix} g_a \\ g_b \\ g_c \end{bmatrix} \quad \text{A.4}$$

$$\mathbf{I}_{dc} = \mathbf{g^T I} \quad \text{A.5}$$

$$\mathbf{I} = \left(\mathbf{g^T}\right)^{-\mathbf{1}} I_{dc} \quad \text{A.6}$$

$$\dot{\mathbf{I}} = \left(\mathbf{g^T}\right)^{-\mathbf{1}} \dot{I}_{dc} - \left(\mathbf{g^T}\right)^{-\mathbf{2}} I_{dc} \quad \text{A.7}$$

$$\mathbf{V} = \mathbf{pg}V_{dc} \quad \text{A.8}$$

$$\left(\mathbf{g^T}\right)^{-\mathbf{1}} \dot{I}_{dc} - \left(\mathbf{g^T}\right)^{-\mathbf{2}} I_{dc} = -\mathbf{L}^{-1}\left(\mathbf{R}\left(\mathbf{g^T}\right)^{-\mathbf{1}} I_{dc} + \mathbf{pg}V_{dc}\right) + \mathbf{L}^{-1}\left(\frac{d}{d\theta_e}\mathbf{L_f}\left(\theta_e\right)\omega_e I_f + \mathbf{L_f}\left(\theta_e\right)\dot{I}_f\right) \quad \text{A.9}$$

A. Salah, *Model-Based Fault Diagnosis and Fault-Tolerant Control*, Wissenschaftliche Reihe Fahrzeugtechnik Universität Stuttgart, https://doi.org/10.1007/978-3-658-50108-2

$$\dot{I}_{dc} = \left(\left(\mathbf{g}^{\mathbf{T}}\right)^{-\mathbf{1}} - \mathbf{L}^{-1}\mathbf{R}\right) I_{dc} - \left(\mathbf{g}^{\mathbf{T}}\right)\mathbf{L}^{-1}\mathbf{pg}V_{dc} + \left(\mathbf{g}^{\mathbf{T}}\right)\mathbf{L}^{-1}\left(\frac{d}{d\theta_e}\mathbf{L_f}\left(\theta_e\right)\omega_e I_f + \mathbf{L_f}\left(\theta_e\right)\dot{I}_f\right) \quad \text{A.10}$$

$$\dot{I}_f = -L_f^{-1}\left(R_f I_f + DV_{dc}\right) + L_f^{-1}\left(\frac{d}{d\theta_e}\mathbf{L_f}\left(\theta_e\right)\omega_e\left(\mathbf{g}^{\mathbf{T}}\right)^{-\mathbf{1}} I_{dc}\right) + L_f^{-1}\left(\mathbf{L_f}\left(\theta_e\right)\left(\left(\mathbf{g}^{\mathbf{T}}\right)^{-\mathbf{1}}\dot{I}_{dc} - \left(\mathbf{g}^{\mathbf{T}}\right)^{-\mathbf{2}} I_{dc}\right)\right) \quad \text{A.11}$$

$$\begin{aligned}\dot{I}_{dc} = &\left(\left(\mathbf{g}^{\mathbf{T}}\right)^{-\mathbf{1}} - \mathbf{L}^{-1}\mathbf{R}\right) I_{dc} - \left(\mathbf{g}^{\mathbf{T}}\right)\mathbf{L}^{-1}\mathbf{pg}V_{dc} \quad \text{A.12}\\ &+ \left(\mathbf{g}^{\mathbf{T}}\right)\mathbf{L}^{-1}\frac{d}{d\theta_e}\mathbf{L_f}\left(\theta_e\right)\omega_e I_f\\ &+ \left(\mathbf{g}^{\mathbf{T}}\right)\mathbf{L}^{-1}\mathbf{L_f}\left(\theta_e\right)\left(-L_f^{-1}\left(R_f I_f + DV_{dc}\right)\right)\\ &+ \left(\mathbf{g}^{\mathbf{T}}\right)\mathbf{L}^{-1}\mathbf{L_f}\left(\theta_e\right)L_f^{-1}\left(\frac{d}{d\theta_e}\mathbf{L_f}\left(\theta_e\right)\omega_e\left(\mathbf{g}^{\mathbf{T}}\right)^{-\mathbf{1}} - \mathbf{L_f}\left(\theta_e\right)\left(\mathbf{g}^{\mathbf{T}}\right)^{-\mathbf{2}}\right) I_{dc}\\ &+ \left(\mathbf{g}^{\mathbf{T}}\right)\mathbf{L}^{-1}\mathbf{L_f}\left(\theta_e\right)L_f^{-1}\,\mathbf{L_f}\left(\theta_e\right)\left(\mathbf{g}^{\mathbf{T}}\right)^{-\mathbf{1}}\dot{I}_{dc}\end{aligned}$$

The Terms:

$$I_{dc} : \left(\left(\mathbf{g}^{\mathbf{T}}\right)^{-\mathbf{1}} - \mathbf{L}^{-1}\mathbf{R}\right) + \left(\mathbf{L}^{-1}\mathbf{L_f}\left(\theta_e\right)L_f^{-1}\left(\frac{d}{d\theta_e}\mathbf{L_f}\left(\theta_e\right)\omega_e - \mathbf{L_f}\left(\theta_e\right)\left(\mathbf{g}^{\mathbf{T}}\right)^{-\mathbf{1}}\right)\right) \quad \text{A.13}$$

$$V_{dc} : -\left(\mathbf{g}^{\mathbf{T}}\right)\mathbf{L}^{-1}\left(\mathbf{pg} - \mathbf{L_f}\left(\theta_e\right)L_f^{-1}D\right) \quad \text{A.14}$$

$$I_f : \left(\mathbf{g}^{\mathbf{T}}\right)\mathbf{L}^{-1}\left(\frac{d}{d\theta_e}\mathbf{L_f}\left(\theta_e\right)\omega_e - \mathbf{L_f}\left(\theta_e\right)L_f^{-1}R_f\right) \quad \text{A.15}$$

$$\dot{I}_{dc} : \left(1 - \mathbf{L}^{-1}L_f^{-1}\right) \quad \text{A.16}$$

assume that the load is R_{dc}

$$V_f \approx DV_{dc} \tag{A.17}$$

$$V_{dc} = I_{dc} R_{dc} \tag{A.18}$$

$$I_{dc} = R_{dc}^{-1} D^{-1} V_f \tag{A.19}$$

$$I_{dc_{\Delta D \approx 0, \Delta Vf \approx 0}} = R_{dc}^{-1} D^{-1} Vf \approx 0 \tag{A.20}$$

$$\begin{aligned} \dot{I}_f = & -L_f{}^{-1}\left(R_f I_f + V_f\right) \\ & + L_f{}^{-1}\left(\frac{d}{d\theta_e}\mathbf{L_f}\left(\theta_e\right)\omega_e\left(\mathbf{g^T}\right)^{-1} - \mathbf{L_f}\left(\theta_e\right)\left(\mathbf{g^T}\right)^{-2}\right)\left(R_{dc}^{-1} D^{-1} V_f\right) \\ & + L_f{}^{-1}\,\mathbf{L_f}\left(\theta_e\right)\left(\mathbf{g^T}\right)^{-1}\left(R_{dc}^{-1} D^{-1} \dot{V}_f\right) \end{aligned} \tag{A.21}$$

$$\begin{aligned} \dot{I}_f = & -L_f{}^{-1} R_f I_f \\ & + L_f{}^{-1}\left(R_{dc}^{-1} D^{-1}\left(\frac{d}{d\theta_e}\mathbf{L_f}\left(\theta_e\right)\omega_e\left(\mathbf{g^T}\right)^{-1} - \mathbf{L_f}\left(\theta_e\right)\left(\mathbf{g^T}\right)^{-2}\right) - 1\right) V_f \end{aligned} \tag{A.22}$$

Assume that θ_e is a small angle

$$\sin\theta_e \approx \theta_e \tag{A.23}$$

$$\cos\theta_e \approx 1 \tag{A.24}$$

$$\cos\left(\theta_e \pm \Phi\right) \approx \cos\Phi \mp \theta_e \sin\Phi \tag{A.25}$$

$$\sin\left(\theta_e \pm \Phi\right) \approx \sin\Phi \pm \theta_e \cos\Phi \tag{A.26}$$

$$\mathbf{L_f}\left(\theta_e\right) = \begin{bmatrix} M\cos\theta_e \\ M\cos\left(\theta_e + \Phi\right) \\ M\cos\left(\theta_e - \Phi\right) \end{bmatrix} \tag{A.27}$$

$$\frac{d}{d\theta_e}\mathbf{L_f}\left(\theta_e\right) = \begin{bmatrix} -M\sin\theta_e \\ -M\sin\left(\theta_e + \Phi\right) \\ M\sin\left(-\theta_e + \Phi\right) \end{bmatrix} \tag{A.28}$$

$$\mathbf{L_f}(\theta_e)\cdot\frac{d}{d\theta_e}\mathbf{L_f}(\theta_e)=\begin{bmatrix}-M^2\sin\theta_e\cos\theta_e\\-M^2\sin(\theta_e+\Phi)\cos(\theta_e+\Phi)\\-M^2\sin(\theta_e-\Phi)\cos(\theta_e-\Phi)\end{bmatrix}\qquad \text{A.29}$$

$$\approx\begin{bmatrix}-M^2\theta_e\\-M^2\left(\sin\Phi\cos\Phi+\theta_e\left(1-2\sin^2\Phi\right)\right)\\-M^2\left(\sin\Phi\cos\Phi+\theta_e\left(1-2\cos^2\Phi\right)\right)\end{bmatrix}$$

$$\mathbf{L_f}(\theta_e)\cdot\frac{d}{d\theta_e}\mathbf{L_f}(\theta_e)\approx\begin{bmatrix}-M^2\theta_e\\-M^2\left(\sin\Phi\cos\Phi+\theta_e\left(1-2\sin^2\Phi\right)\right)\\-M^2\left(\sin\Phi\cos\Phi+\theta_e\left(1-2\cos^2\Phi\right)\right)\end{bmatrix}\qquad \text{A.30}$$

$$=\begin{bmatrix}-M^2\theta_e\\-M^2\left(-\frac{\theta_e}{2}-0.433\right)\\-M^2\left(\frac{\theta_e}{2}-0.433\right)\end{bmatrix}$$

$$=-M^2\left(\theta_e-0.866\right)$$

$$=\mathbf{d}$$

$$\mathbf{L_f}(\theta_e)=\begin{bmatrix}M\cos\theta_e\\M\cos(\theta_e+\Phi)\\M\cos(\theta_e-\Phi)\end{bmatrix}\qquad \text{A.31}$$

$$\approx\begin{bmatrix}M\\M\left(\cos\Phi-\theta_e\sin\Phi\right)\\M\left(\cos\Phi+\theta_e\sin\Phi\right)\end{bmatrix}$$

$$\frac{d}{d\theta_e}\mathbf{L_f}(\theta_e)\approx\begin{bmatrix}0\\M\left(0-\sin\Phi\right)\\M\left(0+\sin\Phi\right)\end{bmatrix}\qquad \text{A.32}$$

$$=\begin{bmatrix}0\\-M\sin\Phi\\+M\sin\Phi\end{bmatrix}=0$$

$$a_0\dot{I}_{dc}=a_1I_{dc}+a_2\omega_e+a_3V_{dc}+a_4I_f\qquad \text{A.33}$$

Substitute A.32 and A.30 in A.12

$$\dot{I}_{dc} : \left(1 - \mathbf{L}^{-1} L_f{}^{-1}\right) = a_0 \qquad \text{A.34}$$

$$I_{dc} : \left(\mathbf{g^T}\right)^{-\mathbf{1}} \left(1 - \mathbf{L}^{-1}\mathbf{R} - \mathbf{L}^{-1} L_f{}^{-1} \mathbf{L_f}^{\mathbf{2}} (\theta_e)\right) = a_1 \Rightarrow \gamma \qquad \text{A.35}$$

$$\omega_e : \mathbf{L}^{-1} L_f{}^{-1} \mathbf{dM} I_{dco} = a_2 \Rightarrow \eta \qquad \text{A.36}$$

$$V_{dc} : -\left(\mathbf{g^T}\right) \mathbf{L}^{-1} \left(\mathbf{pg} - \mathbf{L_f} (\theta_e) L_f{}^{-1} D\right) = a_3 \Rightarrow \lambda \qquad \text{A.37}$$

$$I_f : -\left(\mathbf{g^T}\right) \mathbf{L}^{-1} \mathbf{L_f} (\theta_e) L_f{}^{-1} R_f = a_4 \Rightarrow \kappa \qquad \text{A.38}$$

$$I_f : -\left(\mathbf{g^T}\right) \mathbf{L}^{-1} \mathbf{L_f} (\theta_e) L_f{}^{-1} R_f = a_4 \Rightarrow \kappa \qquad \text{A.39}$$

$$\dot{I}_{dc} = -\gamma\, I_{dc} + \eta \omega_e + \kappa\, I_f - \lambda\, V_{dc} \qquad \text{A.40}$$

$$\gamma = \frac{\mathbf{L} L_f - \mathbf{L_f}^{\mathbf{2}} (\theta_e) - L_f \mathbf{R}}{\mathbf{g^T} (1 - \mathbf{L} L_f)} = \frac{\Omega^2}{\Omega^2} = 1 \qquad \text{A.41}$$

$$\eta = \frac{\mathbf{dM} I_{dco}}{\mathbf{L} L_f - 1} = \Omega.A = V\lambda \qquad \text{A.42}$$

$$= \frac{(\mathbf{g^T}) (\mathbf{L_f} (\theta_e) D - \mathbf{pg} L_f)}{1 - \mathbf{L} L_f} \frac{\Omega}{\Omega^2} = \frac{1}{\Omega} \qquad \text{A.43}$$

$$\kappa = \frac{(\mathbf{g^T}) \mathbf{L_f} (\theta_e) R_f}{1 - \mathbf{L} L_f} = \frac{\Omega^2}{\Omega^2} = 1 \qquad \text{A.44}$$

$$\dot{I}_f = -L_f{}^{-1} R_f I_f + L_f{}^{-1} \left(-R_{dc}^{-1} D^{-1}\, \mathbf{L_f} (\theta_e) \left(\mathbf{g^T}\right)^{-\mathbf{2}} - 1\right) V_f \qquad \text{A.45}$$

$$= -\alpha\, I_f + \beta V_f$$

$$\alpha = \frac{R_f}{L_f} \qquad \text{A.46}$$

$$\beta = \frac{\left(-R_{dc}^{-1} D^{-1}\, \mathbf{L_f} (\theta_e) (\mathbf{g^T})^{-\mathbf{2}} - 1\right)}{L_f} \qquad \text{A.47}$$

B. Parameter Deviations at Several Operating points

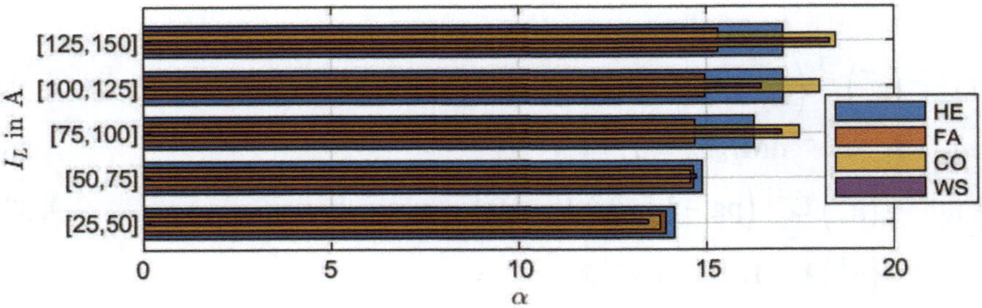

Figure B.1: Deviation in α Caused by Different Faults at $N = 4500$ rpm

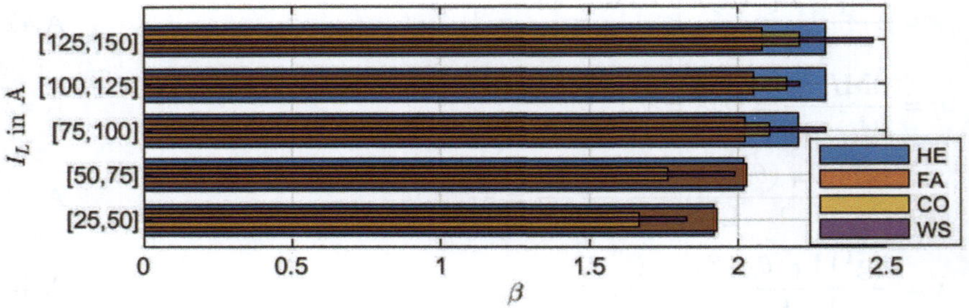

Figure B.2: Deviation in β Caused by Different Faults at $N = 4500$ rpm

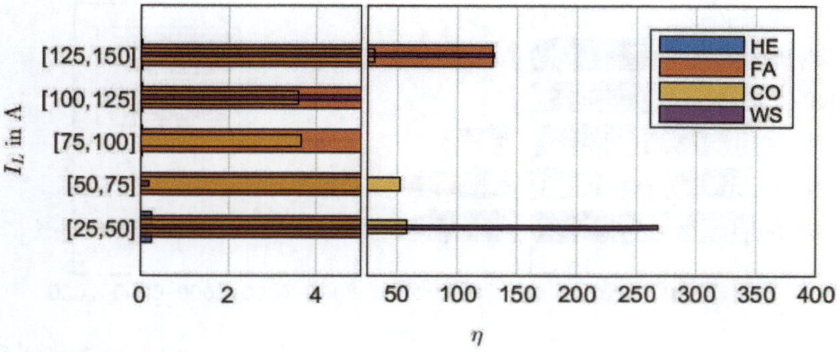

Figure B.3: Deviation in η Caused by Different Faults at $N = 4500$ rpm

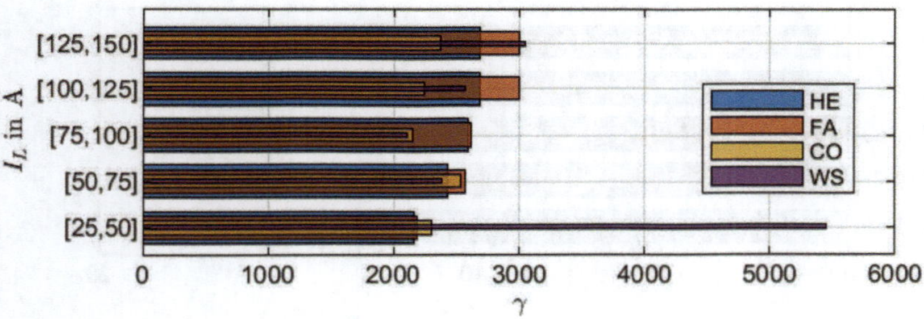

Figure B.4: Deviation in γ Caused by Different Faults at $N = 4500$ rpm

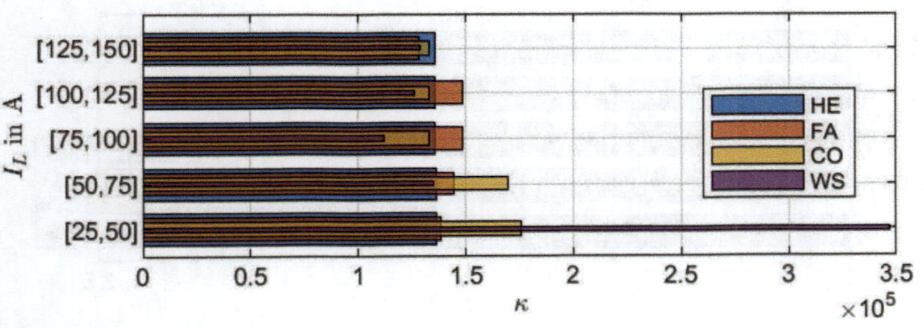

Figure B.5: Deviation in κ Caused by Different Faults at $N = 4500$ rpm

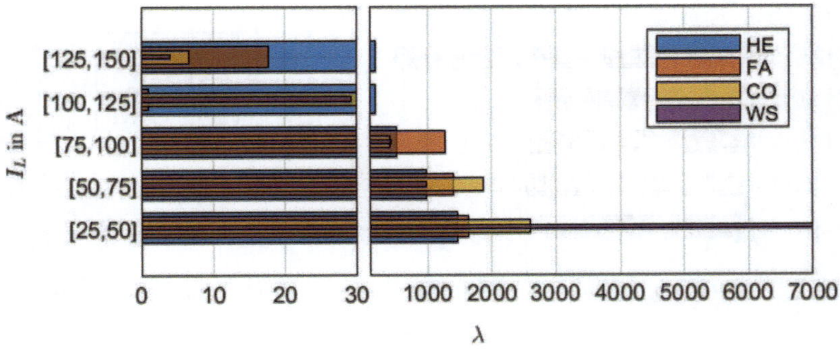

Figure B.6: Deviation in λ Caused by Different Faults at $N = 4500$ rpm

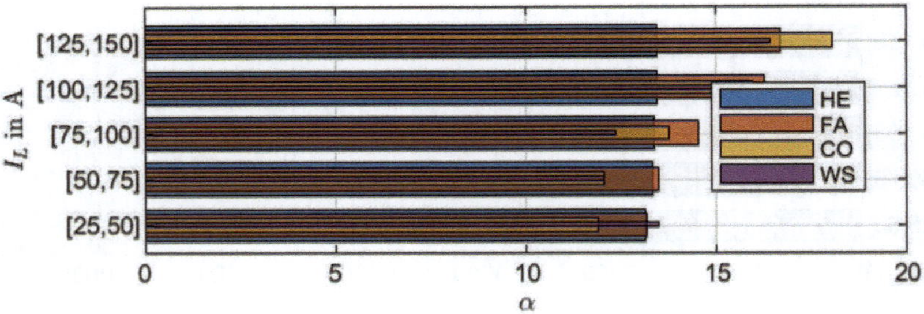

Figure B.7: Deviation in α Caused by Different Faults at $N = 6000$ rpm

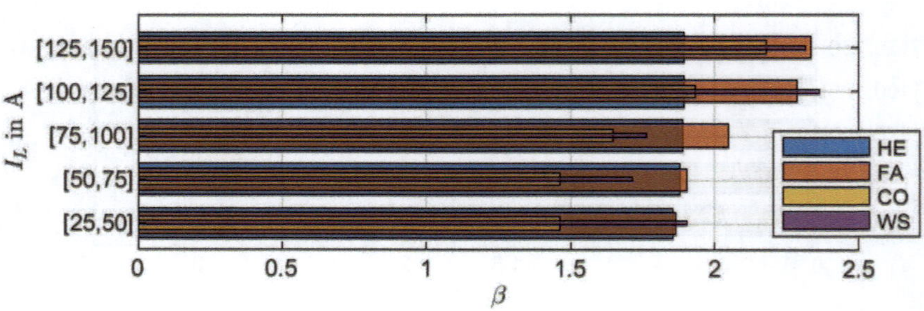

Figure B.8: Deviation in β Caused by Different Faults at $N = 6000$ rpm

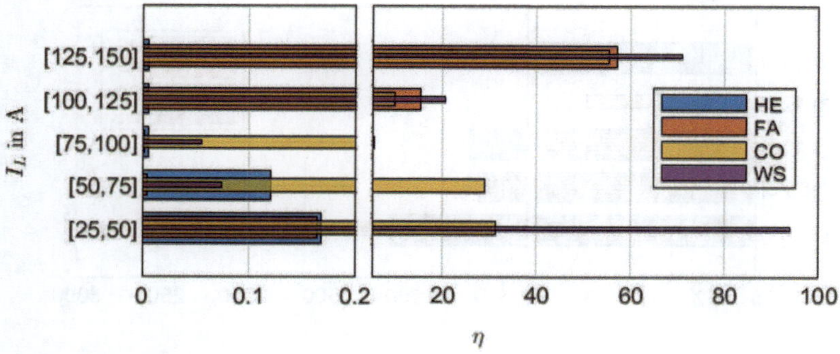

Figure B.9: Deviation in η Caused by Different Faults at $N = 6000$ rpm

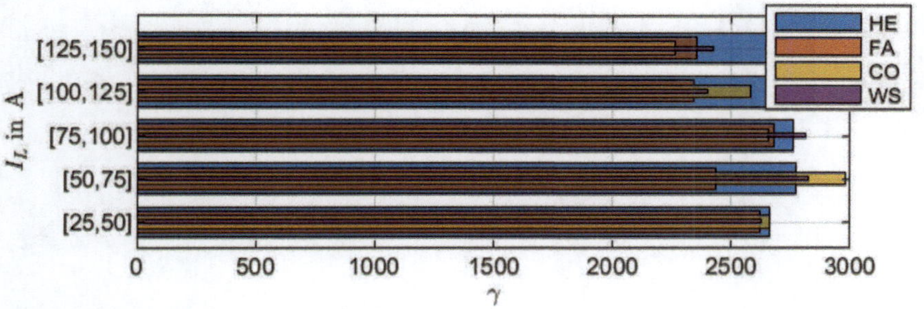

Figure B.10: Deviation in γ Caused by Different Faults at $N = 6000$ rpm

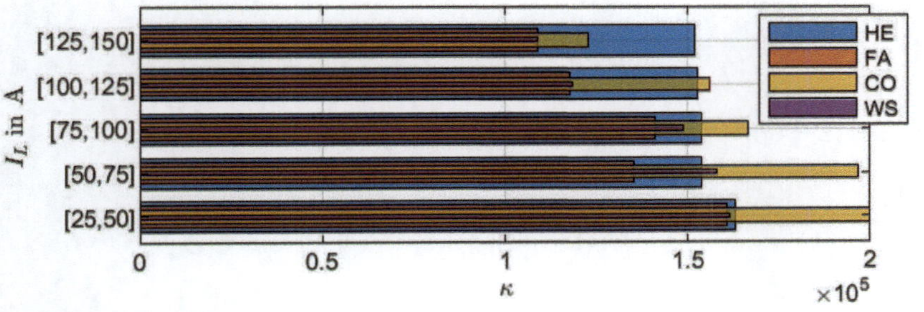

Figure B.11: Deviation in κ Caused by Different Faults at $N = 6000$ rpm

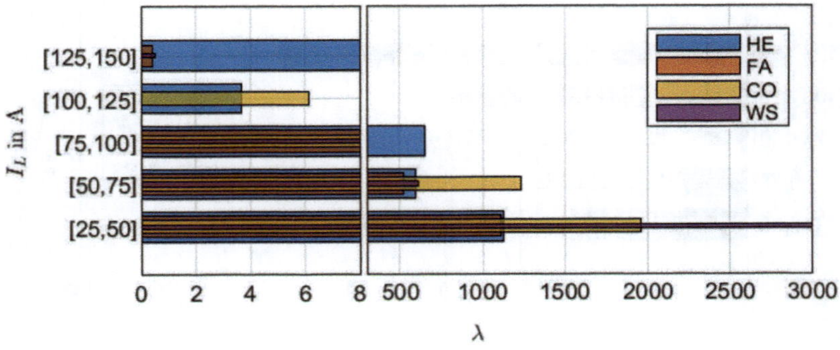

Figure B.12: Deviation in λ Caused by Different Faults at $N = 6000\,\mathrm{rpm}$

C. Proof of Stability of Model Reference Adaptive Control

Consider: $V(x) > 0, V(0) = 0, V$ is an energy function and is radially unbounded

1. $\dot{V}(X) \le 0$ — Laypunov Stable (x bounded)

   To get more information about the stability → Barbalat's Lemma →La Salle Invariance

2. $\dot{V}(X) < 0$ — Asymptotically Stable $\lim_{t\to\infty} x(t) = 0$

3. $\dot{V}(X) < -\alpha V$ — Exponentially Stable $(V \le V(x(0)) \cdot e^{-\alpha t})$ Convergence with α rate

4. $\dot{V}(X) < -\alpha V + \beta$ — Boundednes $\left(V \le V(x(0)) \cdot e^{-\alpha t} + \frac{\beta}{\alpha}\right)$ Comparison Principle

Foundations of Stability in Adaptive Control

Assume A_r is Hurwitz. For $R > 0$, $A_r P + P A_r + R = 0$ holds with a unique $P > 0$. **General proof technique:**

$$\dot{e}(t) = A_r\, e(t) - B\Lambda\tilde{W}^T(t)\Theta(\cdot) + B\Lambda\varepsilon(\cdot) \qquad \text{C.1}$$

Where $\dot{e} \in \mathbb{R}^{n\times 1}$, $A_r \in \mathbb{R}^{n\times n}$,$e \in \mathbb{R}^{n\times 1}$, $B \in \mathbb{R}^{n\times m}$, $\Lambda \in \mathbb{R}^{m\times m}$, $\tilde{W}^T(t) \in \mathbb{R}^{s\times m}$,$\Theta \in \mathbb{R}^{s\times 1}$, and $\varepsilon \in \mathbb{R}^{m\times 1}$, $e(0) = e_o$

$$\dot{\tilde{W}}(t) = \gamma\left(\theta(\cdot)e^T(t)PB - \sigma\tilde{W}(t)\right) - \dot{W}(t) \qquad \text{C.2}$$

Where:

- $\tilde{W}(0) = \tilde{W}_o$
- $\gamma > 0$ Adaptation rate
- $\sigma > 0$ Sigma modification gain

$$\tilde{W}(t) \triangleq \hat{W}(t) - W(t). \qquad \text{C.3}$$

$\dot{e}(t) = A_r\, e(t) - B\Lambda\tilde{W}^T(t)\Theta(\cdot) + B\Lambda\varepsilon(\cdot), \quad e(0) = e_0.$

$\dot{\tilde{W}}(t) = \gamma\left(\theta(\cdot)e^T(t)PB - \sigma\tilde{W}(t)\right) - \dot{W}(t), \tilde{W}(0) = \tilde{W}_0$

Assumptions

1. A_r is Hurwitz $\Longrightarrow A_r P + P A_r + R = 0, \quad R > 0, \quad P > 0.$
2. If $e(t)$ is bounded, then $\Theta(\cdot)$ bounded.

3. $W(t)$ and $\dot{W}(t)$ are bounded:

$$\| W(t) \|_2 \leq \overline{W}, \quad \| \dot{W}(t) \|_2 \leq \overline{\dot{W}}, \overline{W}(t) > 0, \overline{\dot{W}} > 0$$

4. $\varepsilon(\cdot)$ is bounded: $\| \varepsilon(\cdot) \|_2 \leq \overline{\varepsilon}, \quad \overline{\varepsilon} > 0$

5. Λ is a positive - definite matrix with a scalar diagonal elements

$$\begin{bmatrix} \Lambda_1 & 0 & 0 \\ 0 & \ddots & 0 \\ 0 & 0 & \Lambda_m \end{bmatrix} \tag{C.4}$$

Lyapunov Candidate Function

$$V(e, \tilde{W}) = e^T Pe + \gamma^{-1} \cdot tr\left(\left(\tilde{W}\Lambda^{\frac{1}{2}}\right)^T \cdot \left(\tilde{W}\Lambda^{\frac{1}{2}}\right)\right) > 0 \tag{C.5}$$

Where P is the solution of the Lyapunov equation. The first term is scalar and the second one is scalar, so the Lyapunov function is mostly scalar.

- Differentiate the first term:

$$\begin{aligned} \frac{d}{dt}\left(e^T Pe\right) &= 2e^2 P\dot{e} \\ &= 2e^T P\left[A_r e - B\Lambda\tilde{W}^T\Theta + B\Lambda\varepsilon\right] \\ &= 2e^T PA_r e - 2e^T PB\Lambda\tilde{W}^T\Theta + 2e^T PB\Lambda\varepsilon \end{aligned} \tag{C.6}$$

$$\begin{aligned} 2e^T PA_r e &= e^T PA_r e + e^T PA_r e \\ &= e^T PA_r e + \left(e^T PA_r e\right)^T \\ &= e^T PA_r e + e^T A_r^T Pe \\ &= e^T\left[A_r^T P + PA_r\right] e \\ &= -e^T Re \end{aligned} \tag{C.7}$$

$$2e^T PB\Lambda\tilde{W}^T\Theta = \underbrace{e^T PB\Lambda}_{x^T}\,\underbrace{\tilde{W}^T\Theta}_{y} \tag{C.8}$$

$$\begin{aligned} x^T y &= tr(yx^T) \\ &= -2e^T PB\Lambda\tilde{W}^T\Theta \\ &= -2tr\left[\tilde{W}^T\Theta e^T PB\Lambda\right] \end{aligned}$$

$$2e^T PB\Lambda\varepsilon \le 2\|e\|_2\,\underbrace{2\|PB\|_2\|\Lambda\|\overline{\varepsilon}}_{d_1} \tag{C.9}$$

$$\frac{d}{dt}\left(e^T Pe\right) \le -e^T Re - 2tr\left[\tilde{W}\Theta e^T PB\Lambda\right] + 2d_1\|e\|_2 \tag{C.10}$$

- Differentiate the second term:

$$\begin{aligned} \frac{d}{dt}\left(\gamma^{-1}tr\left[\left(\tilde{W}\Lambda^{\frac{1}{2}}\right)^T\left(\tilde{W}\Lambda^{\frac{1}{2}}\right)\right]\right) &= \frac{d}{dt}\left(\gamma^{-1}tr\left[\lambda^{\frac{1}{2}}\tilde{W}^T\tilde{W}\Lambda^{\frac{1}{2}}\right]\right) \\ &= \frac{d}{dt}\left(\gamma^{-1}tr\left[\tilde{W}^T\tilde{W}\Lambda\right]\right) \\ &= 2\gamma^{-1}tr\left[\tilde{W}^T\dot{\tilde{W}}\Lambda\right] \\ &= 2\gamma^{-1}tr\left[\tilde{W}^T\left[\gamma\left(\theta e^T PB - \sigma\hat{W}\right) - \dot{W}\right]\Lambda\right] \end{aligned} \tag{C.11}$$

$$= 2tr\left[\tilde{W}^T\theta e^T PB\Lambda\right] - 2\sigma tr\left[\tilde{W}^T\hat{W}\right]\Lambda - \gamma^{-1}tr\left[\tilde{W}^T\dot{W}\Lambda\right] \tag{C.12}$$

$$\begin{aligned} -2\sigma tr\left[\tilde{W}^T\hat{W}\right]\Lambda &= 2\sigma tr\left[\tilde{W}^T\left(\tilde{W} + W\right)\right]\Lambda \\ &= -2\sigma tr\left[\tilde{W}^T\tilde{W}\Lambda\right] - 2\sigma tr\left[\tilde{W}^T W\Lambda\right] \end{aligned} \tag{C.13}$$

$$= 2tr\left[\tilde{W}^T\theta e^T PB\Lambda\right] - 2\sigma tr\left[\tilde{W}^T\tilde{W}\Lambda\right] - 2tr\left[\tilde{W}^T\underbrace{\left[\sigma W + \gamma^{-1}\dot{W}\right]}_{M}\right]\Lambda \tag{C.14}$$

Recall Lyapunov Function Candidate

$$V\left(e, \tilde{W}\right) = e^T Pe + \gamma^{-1} tr\left[\left(\tilde{W}^T \Lambda^{\frac{1}{2}}\right)^T \cdot \left(\tilde{W} \Lambda^{\frac{1}{2}}\right)\right] > 0 \qquad \text{C.15}$$

$$\begin{aligned}\dot{V} \leq -e^T Re - 2tr\left[\tilde{W}\Theta e^T PB\Lambda\right] + 2d_1\|e\|_2 + 2tr\left[\tilde{W}\Theta e^T PB\Lambda\right] \\ - 2\sigma tr\left[\tilde{W}^T\tilde{W}\Lambda - 2tr\tilde{W}^T M\Lambda\right]\end{aligned} \qquad \text{C.16}$$

- If $\varepsilon(\cdot)$, $\dot{W(t)}$ and σ are zero $\Longrightarrow$ $\dot{V} = -e^T Re \leq 0$ $\Longrightarrow$ Lyapunov Stability
- To apply Barbalt's Lemma $\Longrightarrow$ check $\ddot{V} = -2e^T R\dot{e}$

$$= -2 \underbrace{e^T}_{\text{bd}} R \left[A_r \underbrace{e}_{\text{bd}} - B\Lambda \underbrace{\tilde{W}^T}_{\text{bd}} \Theta(\cdot) \right] \Longrightarrow \ddot{V} \text{ bounded}$$

From Barbalt's Lemma $\Longrightarrow$ $\lim_{t\to\infty} \dot{V} = 0$ $\Longrightarrow$ $\lim_{t\to\infty} e(t) = 0$

- Young's Inequality: $2xy \leq \mu x^2 + \frac{1}{\mu} y^2, \;\; \mu > 0$

$$2d_1\|e\|_2 \leq \mu_1\|e\|_2^2 + \frac{1}{\mu_1}d_1^2, \;\; \mu_1 > 0 \qquad \text{C.17}$$

$$\begin{aligned}-2tr\left[\tilde{W}^T M\Lambda\right] &\leq \mu_2 tr\left[\tilde{W}^T\tilde{W}\Lambda\right] + \frac{1}{\mu_2} tr\left[M^T M\Lambda\right] \\ &\leq \mu_2 tr\left[\tilde{W}^T\tilde{W}\Lambda\right] + \frac{1}{\mu_2} d_2\end{aligned}$$

$$tr\left[M^T M\lambda\right] \leq d_2 \qquad \text{C.18}$$

$$\dot{V} \leq -\underbrace{(\lambda(R) - \mu_1)}_{d_3 > 0}\|e\|_2^2 - \underbrace{(2\sigma - \mu_2)}_{d_4 > 0} tr\left[\tilde{W}^T\tilde{W}\Lambda\right] + d_0 \qquad \text{C.19}$$

$$d_0 \triangleq \frac{1}{\mu_1}d_1^2 + \frac{1}{\mu_2}d_2 \qquad \text{C.20}$$

- Now we should write this in $\dot{V} = -b_0 V + d_0$ in order to use comparison principle!
- $\|e\|_2^2 = e^T e \geq \frac{1}{\lambda(p) e^T Pe}$

- Equation C.19 becomes:

$$\leq -\frac{d_3}{\overline{\overline{\lambda}}} e^T Pe - d_4 tr\left[\left(\tilde{W}\Lambda^{\frac{1}{2}}\right)^T\left(\tilde{W}\Lambda^{\frac{1}{2}}\right)\right] + d_0 \tag{C.21}$$

$$\leq -b_0\left(e^T Pe + tr\left[\left(\tilde{W}\Lambda^{\frac{1}{2}}\right)^T\left(\tilde{W}\Lambda^{\frac{1}{2}}\right)\right]\right) + d_0$$

$$b_0 \triangleq min\frac{d_3}{\overline{\overline{\lambda}}}(p), d_4$$

$\Rightarrow V\left(e, \tilde{(}W)\right) \leq V_0 e^{-b_0 t} + \frac{d_0}{b_0}, V_0 \triangleq V\left(e_0, \tilde{(}W_0)\right)$

$\left(e, \tilde{(}W)\right)$ are bounded, and large γ and small $\overline{\varepsilon}$ tend to make $\frac{d_0}{b_0}$ small and (σ should not be large as well in practice)

MIX
Papier aus verantwortungsvollen Quellen
Paper from responsible sources
FSC® C105338

If you have any concerns about our products,
you can contact us on
ProductSafety@springernature.com

In case Publisher is established outside the EU,
the EU authorized representative is:
Springer Nature Customer Service Center GmbH
Europaplatz 3, 69115 Heidelberg, Germany

Printed by Libri Plureos GmbH
in Hamburg, Germany